FOR HEALTH OR PROFIT?

FOR HEALTH OR PROFIT?

Medicine, the Pharmaceutical Industry,
and the State in New Zealand

Edited by
PETER DAVIS

Auckland
OXFORD UNIVERSITY PRESS
Oxford   New York   Melbourne

Oxford University Press

Oxford University Press, Walton Street, Oxford OX2 6DP

OXFORD NEW YORK TORONTO
DELHI BOMBAY CALCUTTA MADRAS KARACHI
PETALING JAYA SINGAPORE HONG KONG TOKYO
NAIROBI DAR ES SALAAM CAPE TOWN
MELBOURNE AUCKLAND
and associated companies in
BERLIN and IBADAN

Oxford is a trade mark of Oxford University Press

First published 1992
© Oxford University Press

© 1992: Chapter 1 (Peter Davis), 2 (Astrid Baker), 3 (Pauline Norris),
4 (Helen Clark), 5 (Neil Pearce), 6 (Phillida Bunkle), 7 (Sandra Coney),
8 (Rodney Jackson and Ichiro Kawachi), 9 (Ichiro Kawachi), 10 (Sandra Coney),
11 (Alex Thomson), 12 (Joel Lexchin and Pauline Norris), 13 (Ichiro Kawachi
and Joel Lexchin), 14 (Ichiro Kawachi), and 15 (Peter Davis).

This book is copyright.
Apart from any fair dealing for the purpose of private study,
research, criticism, or review, as permitted under the
Copyright Act, no part may be reproduced by any process
without the prior permission
of the Oxford University Press.

ISBN 0 19 558243 8

Cover designed by Hal Chapman
Photoset in Baskerville by Typocrafters Ltd
and printed in New Zealand
Published by Oxford University Press
1A Matai Road, Greenlane, Auckland 5, New Zealand

Contents

Contributors	vii
Acknowledgements	ix

Introduction

Pharmaceuticals and Public Policy *Peter Davis*	1

The Rules of the Game

Setting the Rules: Pharmaceutical Benefits and the Welfare State *Astrid Baker*	18
The Changing Role of Pharmacists and the Distribution of Pharmaccuticals in New Zealand *Pauline Norris*	36
Pharmaceutical Costs and Regulation: From the Minister's Desk *Helen Clark*	53

Science and Safety

Adverse Reactions: The Fenoterol Saga *Neil Pearce*	75
Withdrawal of the Copper 7: The Regulatory Framework and the Politics of Population Control *Phillida Bunkle*	98
A Living Laboratory: The New Zealand Connection in the Marketing of Depo-Provera *Sandra Coney*	119

Mass Medication

The Mass Treatment Trap: Hypertension and Hypercholesterolaemia *Rodney Jackson and Ichiro Kawachi*	144

The Pressure to Treat: Doctors, the Pharmaceutical Industry,
   and the Mass Treatment of Hypertension   162
Ichiro Kawachi

The Exploitation of Fear: Hormone Replacement Therapy and
   the Menopausal Woman   179
Sandra Coney

The Golden Triangle

Choosing a Remedy   208
Alex Thomson

Patents and Generics: A Campaign Diary   229
Joel Lexchin and Pauline Norris

Doctors and the Drug Industry: Therapeutic Information or
   Pharmaceutical Promotion?   245
Ichiro Kawachi and Joel Lexchin

Where's the Bite? — Voluntary Regulation of Pharmaceutical
   Advertising and Promotion   269
Ichiro Kawachi

Conclusion

A Guide to Practice   288
Peter Davis

Index   294

Contributors

Astrid Baker is a lecturer in the Department of Management Systems in the Business Studies Faculty at Massey University. She has a long-term interest in the history of international business organizations and their relationships with national governments. At present she is researching the history of the State provision of medicines in New Zealand.

Phillida Bunkle is a senior lecturer in Women's Studies at Victoria University of Wellington. A graduate of Smith College, she has written on aspects of women's health for more than 15 years, and her essays have been published in *Second Opinion* (OUP, 1988). With Sandra Coney she wrote 'An Unfortunate Experiment', and she has a continuing interest in the issues addressed by the Cartwright Inquiry.

Helen Clark is MP for Mt Albert and Deputy Leader of the Opposition. She completed a MA with first-class honours and lectured in Political Studies at the University of Auckland before entering Parliament in 1981. She was Minister of Health in the Fourth Labour Government from January 1989 to November 1990, and has continued as Opposition spokesperson in that portfolio.

Sandra Coney is a consumer advocate in the area of women's health, Director of the women's health group Fertility Action, and author of several books, including *The Unfortunate Experiment* and *The Menopause Industry*.

Peter Davis is a senior lecturer in Medical Sociology in the Department of Community Health at the University of Auckland Medical School. He trained in sociology and statistics at the London School of Economics and took a doctorate in community health at the University of Auckland. He has published in the sociology of dentistry and in class and stratification studies. His current interests are in health policy and health services research.

Rodney Jackson is a senior lecturer in Epidemiology in the Department of Community Health at the University of Auckland. His research interests have been cardiovascular and asthma epidemiology. He is particularly interested in the public-health and economic implications of large-scale pharmaceutical and behavioural interventions for treating and preventing these diseases.

Ichiro Kawachi is a lecturer in epidemiology in the Department of Community Health at the Wellington School of Medicine. His research interests include cardiovascular disease, the microeconomic evaluation of health care, and the regulation of pharmaceutical and tobacco products. He is currently on a fellowship at the Channing Laboratory, Harvard Medical School.

Joel Lexchin is a member of the Active Staff, Department of Emergency Medicine, Toronto Western Hospital. He has written extensively on pharmaceuticals and the pharmaceutical industry in both Canada and New Zealand.

Pauline Norris is completing a doctorate in the Department of Sociology at Victoria University of Wellington on the changing role of pharmacists. She is particularly interested in the impact of technology and government policy on the occupation of pharmacy, and the development of clinical pharmacy.

Neil Pearce holds a Senior Research Fellowship funded by the Health Research Council in the Department of Medicine at the Wellington School of Medicine. His major research interests are in occupational and environmental health epidemiology and asthma.

Alex Thomson is a senior lecturer in General Practice at the University of Auckland and Chairman of the Research Committee of the Royal New Zealand College of General Practitioners.

Acknowledgements

The idea for this book originated with Joel Lexchin, a Canadian doctor who has written extensively on pharmaceuticals and the pharmaceutical industry both in New Zealand and in his own country. Joel compiled a book-length manuscript on the pharmaceutical industry in New Zealand while visiting and working here in the mid 1980s. The title theme and idea for this book, therefore, are his, while the execution and development are mine. Joel has continued to take an interest in the project, with encouragement, helpful advice, and two co-authored chapters.

A second person who has had a lot to do with the genesis of this book is Ichiro Kawachi, a New Zealand epidemiologist who helped organize the Public Health Association workshop on pharmaceutical promotion in 1990 and who edited its proceedings. Ichiro's interest in this also has been a continuing one, a fact that is reflected in the four chapters he has contributed to the book, as sole or co-author.

My thanks to all the authors for their professional approach to the task, and to Robert Beaglehole for his encouragement and for reading chunks of text at short notice. Appreciation is due also to Anne French, Oxford's publisher, for taking up the challenge of a potentially controversial book with such energy, commitment, and continuing good humour.

Two of the contributions — Phillida Bunkle's chapter on the Copper 7 and Sandra Coney's on Depo-Provera — draw on previously collated material, some of which has been published in the *Listener*. Guy Scott of W. Guy Scott and Associates of Wellington supplied the first figure in the introductory chapter, while George Baxter of the Audio-Visual Unit at the Medical School developed the model of the pharmaceutical system that appears in that chapter as well. George also drew up the diagram of linkages used in Pauline Norris's chapter. The figures in the chapter by Rod Jackson and Ichiro Kawachi are adapted from the literature.

Finally, I think we all must be grateful to the pharmaceutical industry for providing such a wondrous domain of study!

Peter Davis
Auckland

Pharmaceuticals and Public Policy

Peter Davis

Introduction

Pharmaceuticals, and the industry that produces and sells them, are a proper subject for attention. Substantial resources are allocated in the purchase of drugs. They are valued, they are effective, and they have strong symbolic significance in our culture. The industry is powerful, well-resourced, and part of a wider network of significant relationships with key institutions in society including the professions, the scientific community, the State, and the media.

This book addresses the various dimensions of this network and places them within a public policy framework. This introductory chapter outlines the principal features of the industry and foreshadows the key policy issues raised in the chapters that follow.

From Craft to Corporate

Drugs — or pharmaceuticals — are an ever present part of life. Most people are taking some form of medication, prescribed or non-prescribed, much of the time, whether it be to control fertility, to manage daily aches and pains, to calm anxieties, to offset the excesses of modern living, or to deal with short- and long-term conditions of a possibly life-threatening nature. In many respects we live in a medicated society; we expect routine access to packaged medications that can powerfully modify our mental and physical functioning. It is one expedient way in which we try to manage the personal inconveniences and disabilities of life.

It is this routine use of medication that distinguishes the position today from the arrangements governing access to powerful substances of this kind in earlier societies. Because of their potential for modifying moods and mental and physical states, pharmacological agents with healing properties have usually been subject to detailed social prescription and control (Geest and Whyte, 1988). The possession of expertise about such substances has, in the past, endowed the holder with special status and power.

Those qualities traditionally have been jealously guarded and hedged about with ritual and symbol. What is different about our society is the industrializaton of an occupational field that was previously the realm of arcane practice by small groups of specialists in the healing, magical or supernatural arts.

The evolution of the manufacture and distribution of medicines from a craft activity to a modern corporate enterprise has been a rapid one. Where once, even within recent memory, the apothecary arduously compounded a limited range of largely natural ingredients to meet the common hurts and afflictions of life, there is now an industry with immense scientific and commercial resources that acts on an international scale. That industry distributes a great range of products, compounded mainly of synthetic ingredients, to meet a plethora of human needs that are being constantly redefined, refined, and expanded (James, 1981: 9-28).

This change in economic and social organization has been striking. And yet, in some very important respects, other features of the institutional and social landscape remain essentially unaffected; they still seem to hark back to an era in which a solo practitioner dispensed empiric compounds to a small and unsophisticated clientele. It is this contrast between the corporate and expansive character of the modern pharmaceutical industry on the one hand, and the somewhat faltering and uncomprehending response of the professions, the scientific community, governments, and popular sentiment on the other, that forms an important theme for this book.

Market Characteristics

The market for pharmaceuticals has distinctive features. It is these characteristics that explain much of the policy interest and public debate that are a hallmark of the scientific, professional, and popular literature in this field. If the pharmaceuticals market operated in a competitive fashion like many other product markets, there would be little cause for debate. But this is no ordinary product and it is traded in no ordinary market.

There are two key characteristics that distinguish pharmaceuticals from other products: their healing properties and their complex and powerful pharmacological effects. Both attributes are important. The first means that pharmaceuticals are endowed with a wider public interest in a way that, say, beauty products or steel ingots are not. The fact that drugs can save lives, relieve suffering, and prevent disability, gives them a special ethical dimension. Access to them cannot be reasonably denied. Furthermore, their use is subject to social judgements of benefit — improvements in health — rather than just intrinsic profitability.

The second attribute means that, in the case of prescription medicines anyway, the average user of pharmaceuticals cannot be expected to select a pharmacological remedy for their needs with any sureness of success or certainty of safety. Typically these judgements are made by a skilled professional. This means that the key figure in the market for pharmaceuticals is not the end-user (that is, the patient) but an intermediary — namely, the medical practitioner and, to a lesser extent, the pharmacist. This raises issues of market efficiency and agency relationships.

Other features of the pharmaceuticals market and the industry are worth noting. Firstly, the world market is large and rapidy growing. Although pharmaceuticals account for less than 2 per cent of total consumption, this is still a vast market and one, moreover, that has been increasing at an average annual rate of over nine per cent, virtually doubling in nominal terms from US$43.6 billion in 1976 to US$94.1 billion in 1985 (World Health Organization (WHO), 1988: 7-15). Secondly, the industry is dominated by a few large transnational corporations. Although there are more than 10,000 companies worldwide, 25 of these firms account for half of all product shipments worldwide. For these major companies almost three-quarters of sales occur outside the home market. American companies are particularly dominant in this market. Thirdly, research and development play a leading role and are a major and growing part of the industry's cost structure, which increased as a percentage of global sales from less than 5 per cent in the 1950s to over 12 per cent in the 1980s. In consequence, a large number of new products are released each year, over 100 annually in the United States for example. The great majority of these products, however, are combinations and duplicates. Probably fewer than a fifth are genuinely new chemical entities, and of these fewer than a third would represent therapeutic improvements of any significance; this means that, worldwide, there are on average about 15 useful new drugs produced each year (WHO, 1988: 31-9).

To anticipate the conclusions of this analysis, it is possible to say that, because of the special character of the product being traded, the pharmaceuticals market lacks many of the features that make other product markets price competitive. This is because the need for medical care is inelastic, because doctors — the major purveyors of the product — are not price-conscious, because patents protect products against market entry, and because, while the tendencies in the market overall are *not* towards concentration and control by a limited number of companies, this cannot be said with the same confidence of the different therapeutic *sub*-markets. Add to this the ethical imperative of access to therapy in time of need, and add also the ready involvement of the State on behalf of its citizens, and one can see that the opportunities for the emergence of genuine price competition are restricted.

The market, nevertheless, remains highly competitive, but in ways that are not reflected in price signals. Medical practitioners are effectively the target market for pharmaceutical companies. Competition, therefore, essentially takes place through innovation and product differentiation, with companies relying on heavy promotion among doctors to create brand loyalty and to achieve high prices in order to recoup their development costs and secure market share. The pattern of competition is one of product differentiation, using patents and brand names backed by heavy promotion to the principal purveyor, the doctor. It is this form of non-price competition targeted at a small professional group that can lead to high prices, waste, duplication, and therapeutically irrational outcomes.

A Critical Tradition

Assessed in the light of the broader public interest, these distinctive features of the pharmaceuticals market have stimulated a lively, frequently critical, literature. Its critical edge issues from the fundamental tension between the industry's goals of profit maximization and society's wider interest in the public health.

These two purposes — profit and health — are not necessarily nor always in conflict, but there is a strong tension between them, and opportunity aplenty for ambiguity and double standards. The industry does have some outstanding scientific and commercial successes to its credit, but they have to be set against a history of *realpolitik* and political and commercial ruthlessness that sits ill beside an oft-expressed ethical purpose.

It is the contrast between purpose and practice that lends a moral edge to the literature. The evidence on this is substantial. One of the most striking studies in this tradition was the analysis by Silverman (1976) of the contrasting claims made to doctors by pharmaceutical companies in key reference works in the United States and Latin America. He took seven major categories of pharmaceuticals and selected preparations that were valuable and widely used, marketed by the same company or an affiliate, had well-established clinical usefulness and known hazards, and were described in key local reference works. The pattern that emerged was consistent; while entries followed a policy of full disclosure in the United States, in the Latin American sources there was a tendency for more indications to be cited and for fewer hazards, contra-indications and side-effects to be listed.

Silverman concluded that there needed to be more rigorous policing of the claims made in such reference works. He recommended that the international medical scientific community should determine which controversies over drug labelling and promotion were 'honest differences of opinion' and which were unacceptable exaggeration of claims and minimization of hazards. The International Federation of Pharmaceutical Manufacturers Associations (IFPMA) has since promulgated a voluntary code of marketing practice (Leisinger, 1989). In a subsequent publication, Silverman, Lee, and Lydecker (1986) note 'gratifying changes' in the information given by the pharmaceutical industry, but assert that a significant problem of international harmonization and standardization remains.

A broader-ranging review of market distortions and the problems they produce is provided by Braithwaite (1984) in his study of corporate practices in the pharmaceutical industry. He outlines the evidence on price-fixing, serious law violation, international bribery and corruption, fraud (in safety testing), and criminal negligence (with unsafe manufactured drugs). In nearly all these areas the industry has a poor record. In addition there is its undoubted success in avoiding open settlement in court for the multiplicity of product liability suits, large and small, including the benchmark case of thalidomide. Despite this unsavoury record, the industry has a high level of profitability.

Again, as in the case of Silverman's research, Braithwaite's objective is not the expression of moral outrage but a hard-headed and realistic appraisal of possible policy options that might curb corporate excess without destroying the essential commercial vigour and scientific creativity of the industry. Braithwaite does not argue for an increase in regulation, but for a mix of strategies including: the international exchange of information; self-regulation and negotiation between regulatory agencies and the industry; increased publicity for violations and full disclosure of scientific and safety data; and other non-market mechanisms. He also advocates various internal institutional changes, such as a greater independence for medical directors and research scientists, the setting up of internal advocates for the public interest, and the acceptance by Chief Executives of responsibility for setting the ethical climate within their companies. In other work on business regulation Braithwaite (1982) has advocated an 'enforced self-regulation' model in which companies write their own rules, enter into explicit agreements with regulatory agencies, and establish internal compliance mechanisms which are policed by these agencies. Lexchin (1990) has cast some doubt on the effectiveness of such measures based on the Canadian experience and has argued instead for the involvement of consumer and other public interest groups in regulatory arrangements. There seems, nevertheless, to be agreement that questionable corporate practices can be curbed without resort to a potentially unwieldy and inflexible bureaucratic apparatus.

The Social Context

Medicines are material objects; they are substances that can be bought and sold as commodities. It is this substantial and tradeable quality that directs our attention to characteristics of the market and its corporate environment. But medicines also exist in particular cultural and social contexts and have culturally determined meanings alongside, and independent of, their biochemical properties. It is this social dimension that has spawned an equally prolific literature, again much of it of a more questioning nature, expressing unease at the increasingly ready reliance on medication in our society, particularly those preparations directed at modifying and controlling mood and behaviour.

The social dimension of pharmaceuticals can be explored at a number of levels. Perhaps the aspect that has been most widely discussed in the scientific literature is the so-called 'placebo effect'; that is, the therapeutic effect of a treatment that flows from its social and psychological attributes rather than its specifically pharmacological effects (Shapiro, 1959). Most studies have recorded a 'placebo effect' of about 30 per cent; this is the kind of therapeutic improvement that can be registered regardless of the kind of intervention. Indeed, Claridge (1970) has argued that there is a 'total drug effect' that is a function of the non-pharmacological properties of the drug, the patient, the practitioner, and even the setting itself. Variations in colour, size, and shape of medications have all been shown to influence patient outcome (Helman, 1990: 170-6).

'Placebo' and 'total drug' effects cannot readily be accommodated within the biomedical paradigm. Indeed, there are a number of respects in which the conventional biomedical view of the prescribing process appears, on closer examination, to be far too simple, relying on a somewhat linear and rationalistic model of investigation, diagnosis, prescription, regime, and pharmaceutical impact that fails to take sufficient account either of external influences — including the industry and the organization of medicine — or of elements of symbolic, ritualistic, and customary behaviour on the part of both patient and practitioner (Mapes, 1980).

An area where the assumptions of the biomedical model of pharmaceuticals have come under particularly close scrutiny has been in the growing and extensive use of psychotropic drugs such as tranquillizers and antidepressants. This has become an issue of particular concern for a number of reasons, including the rapid growth in use in the 1960s and 1970s, the evidence of long-term dependence among some users, the concentration of use in particular population groups such as women, the diagnostic uncertainty associated with psychiatric conditions in general practice, the impact of use on social functioning (such as depression), and the unease about the possible role of the pharmaceutical industry in influencing prescribing patterns and in medicalizing aspects of social life (Mapes, 1980).

In many respects psychotropics provide a case study — albeit in sharp relief — of the larger picture; the advent of these drugs, their increased use and the subsequent controversies, illustrate, perhaps more vividly than in other cases, the relationship of pharmaceuticals to complex changes in social life, medical practice, and social values (Lennard and Cooperstock, 1980). At one level, of course, one has to take account of the scientific and technological developments that made the emergence of these drugs possible. Developments in biochemistry, pharmacology, physiology, and related sciences, together with substantial economic investment and high profitability, have fuelled the process of scientific creativity in pharmaceuticals. Underpinning this intense application of intellectual and material resources is a modern faith in the ultimate ability of science and technology to address and resolve most real-life problems. This is the guiding ethic of the pharmaceutical paradigm.

While technical developments and the modern scientific ethic have paved the way for pharmaceutical breakthroughs such as the psychotropics, it is what Lennard and Cooperstock (1980: 74-6) term 'medicalization', or the application of the medical model to an ever wider range of problems, that is the crucial dynamic to the process. Again, a complex mixture of technical and social values is involved. The fundamental technical assumption, an assumption that underlies medical practice, and that increasingly permeates popular conceptions, is that of biological 'reductionism'. This is the idea that human behaviour, and a growing range of social deviance, is best accounted for in terms of the biological fundamentals of cellular and molecular functioning. Allied to this is a perception of the human body — including the mind — as a machine with relatively interchangeable and standardized components that are responsive to intervention. These are the two core, working scientific assumptions of the medical model. The great potency of this model — and the factor that has encouraged the coining of the term 'medicalization' — has been its association with wider social values and perceptions, particularly the high, even supreme, valuation placed on health in our society, the increasing acceptance of medicine as a moral arbiter, the emergence of the doctor as the all-purpose healer, and the growing legitimacy conferred on interventions applying health definitions and resources to social problems.

The example of psychotropics is illuminating because it illustrates the extension of pharmaceuticals into areas of social intervention and behaviour modification. While the conditions to which psychotropics are applied may be quite specific — anxiety, depression, and other minor psychiatric episodes — the principle being exemplified is of much wider currency; namely, the close and symbiotic relationship between pharmaceutical therapy and the medical model, and the way in which this subtly transforms social 'problems' into individual and personal 'solutions'.

While this insight may be stimulated by the specifics of psychotropics, it can be applied more broadly. Essentially, pharmaceuticals provide a set of solutions that are pharmacological, that are individualized, that are not disruptive of wider social arrangements, and that are highly marketable. These characteristics stand in sharp contrast to many non-pharmacological interventions, both individual and collective. The sharpest contrasts arise with population-based public health measures. This can be seen most starkly in the Third World where basic problems such as poverty and poor hygiene, nutrition, and housing defy any specifically pharmacological solution to the immense burden of ill-health. Yet, tackling these problems would challenge existing power structures and would require painful and controversial social adjustments. In consequence, the underlying patterns of ill-health persist.

Such ethical and policy contrasts are rarely presented so clearly in industrialized societies. Nevertheless, the same principles can be seen to operate. Non-pharmacological solutions at both individual and collective (or population) levels require a greater mobilization of social and political resources. They also frequently involve challenging established practices and institutions and vested interests. By contrast, the pharmacological approach is socially harmonious; it has the attraction of usually leading to the immediate relief of troubling symptoms for individuals, without requiring a reorganization of lifestyle or disruption of wider social and political structures.

The Policy Environment in New Zealand

As in almost every other area of national life, the policy context of pharmaceuticals and the pharmaceutical industry in New Zealand has witnessed unprecedented scrutiny, debate, and change in the period from the mid-1980s. A system that, like many other institutions in New Zealand, had been in what virtually amounted to a 'steady state' for nearly half a century, was rudely jolted by the deregulation of the economy and the reforms of the public sector instituted by the Fourth Labour Government (Holland and Boston, 1990).

A series of events, including the emergence from a price freeze and a devaluation of the currency in 1984, together with the lifting of price control over medicines in 1986 (Sutton, 1988), were associated with an unprecedented increase in pharmaceutical expenditure. In nominal terms that expenditure doubled over the decade to $500 million and its proportion of Vote: Health increased from 10 to 15 per cent (Hepenstall, 1989). The policy 'crunch' can be discerned in the graph outlined below. While the percentage of Government expenditure devoted to health has remained relatively constant in recent years, figure 1 shows that the long-term trend in the per capita cost of pharmaceutical benefits has been steadily to increase.

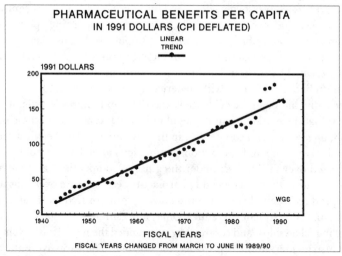

Figure 1

The Government instituted a number of investigations on the health sector, of which two in particular focused on pharmaceuticals. One addressed the removal of medicines from price control (Coopers and Lybrand, 1986) and the other considered the area within the context of health benefits (Scott et al., 1986). Both reports cited evidence that New Zealand was paying much higher prices for its pharmaceuticals than were comparable countries — up to 37 per cent more than Australia, for example — and both advocated the use of the State's powers of monopsony (that is, as sole purchaser) to strike a harder bargain with the industry in negotiating the price of items covered by the pharmaceutical benefits scheme.

The industry, for its part, was also very active in this period. Firstly, it attempted to reposition itself in the public view by, for example, changing the name of its industry organization from the Pharmaceutical Manufacturers Association to the Researched Medicines Industry (RMI). It also launched a media campaign, involving television advertisements, emphasizing the industry's scientific base and its contribution to health advances. Apart from repositioning itself in the public regard and in its relationship with the medical profession, this move also placed the industry in a stronger position to mount future campaigns against the threatened incursion into the New Zealand market of generics (unpatented and unbranded medicines). This was a textbook move in the category of what James (1981: 112-20) has called 'external offensive strategies' against generic drugs. Substrategies, described by James, and readily recognizable in New Zealand over this period, have been 'support[ing] physicians' demands for prescribing autonomy', 'support[ing] consumer groups' demands for better health care', 'induc[ing] prescribing through brand loyalty created by differentiated product', and 'persuad[ing] customers of quality/reliability hazards with generics' (James, 1981: 126).

Secondly, although the evidence is anecdotal and impressionistic, it does appear that the marketing strategies of individual companies became more lavish, aggressive, and competitive in the newly deregulated environment: several advertising and promotional campaigns aimed at the public were conducted over this period; under the guise of competitions and the like, inducements both to doctors and pharmacists became more overt, disputes over product claims between companies were more frequent; there was substantial investment in educational foundations, scholarships, and travelling fellowships; and controversy surrounded the more lavish examples of industry entertainment provided for doctors. The excessive nature of many of these marketing and promotional practices was summarized and highlighted at a workshop convened by the Public Health Association (Kawachi, 1990).

Finally, growing politicization and polarization seemed to mark the industry's relations with the State and with various public interest groups. This reached its height in a more or less explicit campaign by the industry in 1990 to rally public opinion against the Government of the day, signalled by the emergence of an irregular publication called 'The Health Report' and culminating in a series of advertisements in the print media apparently

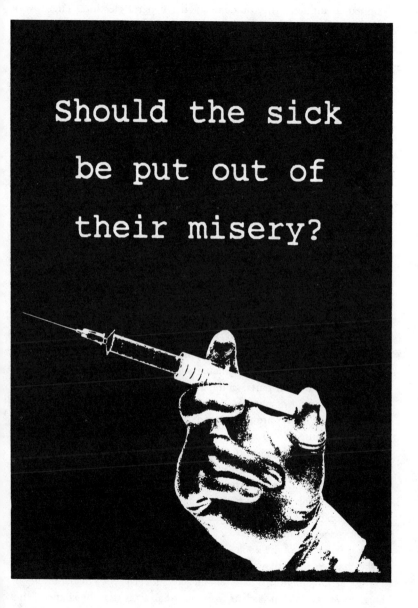

One of a series of four newspaper advertisements run by the Researched Medicines Industry in the lead-up to the 1990 general election.

designed to influence the outcome of the general election. There were several campaigns over this period to gain listing on the Drug Tariff for particular drugs, which inevitably involved conflict with the Government, and there were numerous disputes over price negotiations resulting in litigation. Conflict with public interest groups also marked this period, with controversy over industry tactics on safety issues and promotional practices; furthermore, while the industry publicly espoused dialogue with the key public interest group in the area, the Public Health Association, it was actually working in private for that organization's demise.

The unsavoury nature of the ethical climate came to a head with the resignation of the RMI's chief executive, Bruce Jenkin, in early August 1991. This shortly followed highly publicized price increases for a number of drugs demanded by the company Glaxo in negotiations with the Health Department. What ensued was a media exchange tinged with moral blackmail as both the company and the Government sought to depict the other party as heartless and uncaring, one for raising prices 200–500 per cent, the other for refusing to pay. Jenkin was quoted as saying at the time of his departure from the RMI: 'I find it hard to promote ideas I don't believe in' (*The Examiner*, 22 August 1991).

Issues and Policies

It was in this policy environment and against this background of unease and concern about the ethical climate that this book was initially conceived. It is organized into four parts. In the first section the institutional framework for the distribution of pharmaceuticals in New Zealand is outlined and some key policy issues are introduced. The next two sections explore two areas of continuing controversy in the field of pharmaceuticals: the issue of safety and the question of the ever-broadening application of pharmacological treatment in the population. The final section addresses the key relationships between the three major participants in the distribution of pharmaceuticals — the industry, the medical profession, and the State. A conclusion draws out the principal themes and suggests possible strategies for policy development.

The first section of the book — The Rules of the Game — sets the scene for the chapters that follow. Astrid Baker outlines the historical background to the arrangements that still largely govern the distribution of pharmaceuticals. What is clear from her analysis is that the shape of this structure is inextricably linked to the history and fortunes of the Welfare State in New Zealand. Other themes are rising costs and the changing

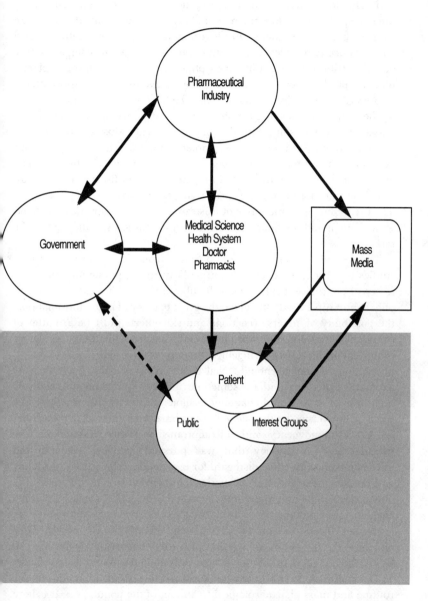

Model of the key relationships in the distribution of pharmaceuticals.

nature of the bureaucratic task (from paying benefits to assessing safety and value for money). Pauline Norris describes the present structure of the system. She examines the role of pharmacy in the context of two case studies about recent change in Government policy. Two features stand out: the uncertain professional status of pharmacy, and the complex nature of the distribution system in which pharmacy is practised, a complexity that can produce perverse and sometimes unwonted changes in patterns of prescribing and dispensing. Finally, Helen Clark considers the 'rules of the game' as they intermesh with the concerns of the State in its bargaining relationships with the pharmaceutical industry. Her analysis suggests that the institutional environment within which such negotiations take place is systematically weighted in the industry's favour. The overriding concerns of the State in circumstances of fiscal stringency are achieving value for money and encouraging rational patterns of prescribing. The predominant theme is one of frustration as the industry deals its hand with acumen and the State finds its chosen instruments insufficient to the task.

The second section, Science and Safety, considers a number of case studies in an area that receives particularly strong public attention, the safety of pharmaceuticals. Given the ubiquity of medication, much of it only lightly supervised, and given the great potency of many formulations, the possibility of adverse reactions and side-effects must be a matter of considerable concern. Since the thalidomide disaster of the 1960s, regulatory requirements have been stringent, although, as these case studies suggest, possibly not stringent enough. Neil Pearce plots a further stage in New Zealand's experience with the sometimes lethal combination of a relatively high level of asthma and changing fashions in prescribing practice. The recent controversy over the safety of the asthma drug fenoterol demonstrates the powerful advocacy and uncompromising stance adopted by the manufacturer, a strategy that was pursued in New Zealand and internationally with a fine disregard for convention. The articles by Coney and Bunkle underscore the international dimension with their case studies on the safety of devices and drugs concerned with fertility control. All three chapters raise questions about the adequacy of arrangements for monitoring safety and efficacy once products have been released onto the market. They also raise doubts about the wisdom of a small country allowing itself to be used as a pioneer in new medical devices and pharmaceutical products.

The contributions in the section entitled Mass Medication address the routine and mass pharmacological treatment of the population. It is here that we can clearly identify the complex intersection between professional and commercial motives that has made pharmaceuticals both a profitable area of investment and an expanding arena of occupational endeavour. The first two chapters analyse what has now become the textbook growth area for pharmaceuticals: namely, the chronic, lifestyle-related conditions, like hypertension or high blood cholesterol, that are widely prevalent in the population, that generate a widespread, if subliminal, level of anxiety, and that exhibit ambiguous indications for clinical intervention. Sandra Coney identifies a related development in an area of interest to women. She uncovers an early stage in the convergence of broadening medical definition and pharmacological remedy with her case study of hormone replacement therapy. In this instance it is a matter of a widely prevalent, life-stage-related condition that lends itself to translation from a 'normal' feature of human development to a medical disorder, thus providing the opportunity for the widespread and prolonged medication of women.

The final section brings to a head the fundamental tripartite relationship of power, influence, and finance that links the industry with the professions — particularly the medical profession — and the State. This raises basic policy issues of control and regulation. This mix of political, legal, financial, and social concerns is exemplified in the area of patents and generics, discussed by Joel Lexchin and Pauline Norris. They outline the ways in which the industry in New Zealand has worked strategically to position public opinion, the relevant professional groups, the bureaucracy, the State, and the legal and scientific framework so as to effectively forestall any significant shift to generic prescribing or to the importation and substitution of brand-name equivalents. Kawachi and Lexchin explore in greater detail the most powerful link in this chain. They show how the industry cultivates its relationship with the medical profession by using a subtle mix of strategies designed to influence prescribing patterns. Against this single-minded and strategic pursuit of the industry's long-term interests, the power of self-regulation, analysed by Alex Thomson and Ichiro Kawachi in their chapters, seems puny indeed. The ethical constraints of the medical profession and the voluntary regulation of advertising and promotion by the industry are shown to be more honoured in the breach than the observance.

Conclusion

It is this contrast between the shrewd use of corporate power on the one hand and the stunted institutional and regulatory framework on the other that remains an enduring impression after examination of these issues. This is *not* to say that the pharmaceutical industry is a social adversary.

It is a vigorous product of advanced, research-based capitalist endeavour. As such it should not be discouraged. At the same time, however, a broader public interest in how drugs are produced, distributed, and used needs to be acknowledged. The pursuit of this broader public interest requires the development of institutional arrangements that can harness the vigour of a global industry to the local health needs and policy concerns of New Zealanders. Suggestions for directions that might be taken towards that objective are discussed in the concluding chapter.

References

Braithwaite, J. (1982) Enforced self regulation: a new strategy for corporate crime control *Michigan Law Review* 80: 1466-507.

Braithwaite, J. (1984) *Corporate Crime in the Pharmaceutical Industry* London, Routledge and Kegan Paul.

Claridge, G. (1970) *Drugs and Human Behaviour* London, Allen Lane.

Coopers and Lybrand Associates (1986) *Removal of Medicines from Price Control* Unpublished report, Wellington, Department of Health.

Geest, S. van der and Whyte, S.R. (eds) (1988) *The Context of Medicines in Developing Countries: Studies in Pharmaceutical Anthropology* Dordrecht, Kluwer Academic Publishers.

Helman, C.G. (1990) *Culture, Health and Illness* London, Wright, 2nd edn.

Hepenstall, B. (1989) Pharmaceutical benefits: setting the level. In Maling, T. (ed) *Drug Prescribing Efficiency* 27-8 Auckland, The Medical Publishing Company Ltd.

Holland, M. and Boston, J. (1990) *The Fourth Labour Government: Politics and Policy in New Zealand* Auckland, Oxford University Press.

James, B.G. (1981) *The Marketing of Generic Drugs* London, Associated Business Press.

Kawachi, K. (ed) *Pharmaceutical Advertising and Promotion — Options for Action* Wellington, Public Health Association of New Zealand.

Leisinger, K.M. (1989) *Poverty, Sickness and Medicines: An Unholy Alliance?* Geneva, International Federation of Pharmaceutical Manufacturers Associations.

Lennard, H.L. and Cooperstock, R. (1980) The social context and functions of tranquilliser prescribing. In Mapes, R. (ed) *Prescribing Practice and Drug Usage* 73-82 London, Croom Helm.

Lexchin, J. (1990) Drug makers and drug regulators: too close for comfort. A study of the Canadian situation *Social Science and Medicine* 31: 1257-63.

Mapes, R. (1980) Introduction. In Mapes, R. (ed) *Prescribing Practice and Drug Usage* 1-10 London, Croom Helm.

Scott, C., Fougere, G. and Marwick, J. (1986) *Choices for Health Care: Report of the Health Benefits Review* Wellington, Health Benefits Review.

Shapiro, A.K. (1959) The placebo effect: a neglected asset in the care of patients *American Journal of Psychiatry* 116: 298-304.

Silverman, M. (1976) *The Drugging of the Americas* Berkeley and Los Angeles, University of California Press.

Silverman, M., Lee, P. and Lydecker, M. (1986) Drug promotion: the Third World revisited *International Journal of Health Services* 16: 659–67.

Sutton, F. (1988) *Trends in Pharmaceutical Benefit Expenditure over the Last Decade* Wellington, Department of Health.

World Health Organization (WHO) (1988) *The World Drug Situation* Geneva, World Health Organization.

Setting the Rules: Pharmaceutical Benefits and the Welfare State

Astrid Baker

Pharmaceutical benefits have been the focus of powerful interests since their introduction in New Zealand in 1941. Drug manufacturers, wholesalers, pharmacists, doctors, and the State all have an interest in providing these benefits. For the State the provision of medicines is an important social welfare commitment. However, this cost to the State is income and profit to drug manufacturers, distributors, and pharmacists. This tension underlies the history of State funding for medicines in New Zealand.

This chapter describes the introduction of pharmaceutical benefits in New Zealand, and examines the history of the key interest groups — the funders, the producers, the prescribers and the dispensers of medicines. The study is set in the context of the emergence of the modern Welfare State and the evolution of a managed economy in New Zealand, both of which had their origins under the first Labour Government. Its end point is the passing of the Commerce Act, 1986, which removed medicines from price control and dismantled complex institutional arrangements of more than 40 years.

Establishing Pharmaceutical Benefits

Before the first Labour Government introduced its far-reaching Social Security Act in 1938, there was no Government subsidy on medicines in New Zealand. The Friendly Societies met the cost of medicines supplied by pharmacists, as one of the benefits provided for their paid-up members (Submissions to National Health and Superannuation Committee, 1938).

The central provisions of the Social Security Act were a series of monetary benefits (for example for children, pensioners, and widows) and the creation of a health service accessible to all. Part III of the Act provided 'to every member of the community full and adequate hospital, sanatoria, medical, pharmaceutical, maternity and other health services' (New Zealand Statutes, 1938). These health services or benefits were to be available to everyone resident in New Zealand, irrespective of their financial position, and were to be administered by the Department of Health. A new Department of Social Security was to administer the monetary benefits.

Pharmaceutical benefits were defined in the Act as the right to claim such medicines, drugs, prescribed materials, and prescribed appliances ordered by any medical practitioner in the course of providing medical services. The Minister of Health would fix the prices to be paid for medicines, and the terms and conditions by which these would be supplied. Contracts with hospitals, doctors, chemists, and others would be negotiated to ensure the introduction and successful operation of the scheme. The Act allowed for the introduction of several classes of benefits step by step, as soon as adequate arrangements could be made. Representatives of the New Zealand Branch of the British Medical Association (BMA), the Friendly Societies' Medical Institute, the Pharmacy Board of New Zealand, the Pharmacy Plan Industrial Committee (PPIC), and the Chemists' Service Guild were all consulted by the Government.

Detailed arrangements for a free general practitioner service, and for the provision of free medicines, were still under negotiation when the rest of the Social Security Act was introduced on 1 April 1939. The cause of the delay was a bitter quarrel between the Labour Government and the New Zealand branch of the BMA over doctors' contracts for medical benefits (Bolitho, 1984; Hanson, 1980). Eventually an agreement was reached in 1941 whereby doctors would charge their patients a fee, and be reimbursed from the Social Security Fund.

Because of the Government's long quarrel with the doctors, pharmaceutical benefits could not be introduced under the appropriate provisions of the Social Security Act. The Government took advantage of its powers under Section 101 of the Act to introduce supplementary benefits. In this way Social Security Pharmaceutical (Supplies) Benefits, which provided for the free supply of pharmaceutical requirements upon the prescription of any registered doctor, were introduced in May 1941. The Minister of Health would now pay from the Social Security Fund for those medicines which doctors chose to prescribe, and which were specified in a Drug Tariff. Registered pharmacists would dispense the doctor's prescription as contractors to the Department of Health. Hospital Boards were also entitled to receive payment for drug requirements supplied to out-patients. There was no prescription charge.

What were the implications of this new legislation? It meant that neither the doctor prescribing, nor the patient consuming the medicine need now be concerned with prices. Pharmacists, too, need not be concerned about the price of prescriptions — a reduction in cost would not increase their volume of business. The higher the cost of ingredients, the greater their gross profit. Drug manufacturers had little to fear from buyer resistance to higher prices. This regime, once established, meant that any levying of charges for prescriptions would be an unpopular and difficult political decision for governments.

The original Drug Tariff published in 1941 set out in two pages the dispensing fees and prices to be paid from the Social Security Fund. The Tariff specified by reference to official pharmacopoeia and, by direct inclusions or exclusions, those medicines which were pharmaceutical benefits and thus to be charged to the Social Security Fund. The Tariff was to be amended at regular intervals. All medicines which appeared in the *British Pharmacopoeia* (1932) and the *British Pharmaceutical Codex* (1934) and their subsequent revisions, were deemed to be included in the Drug Tariff, and to be 'pharmaceutical requirements'. In order to qualify for inclusion a medicine had to be prepared according to the directions set out in these official publications. No official formulary was mentioned at this stage.

Limits were imposed on the quantities of drugs that could be supplied in any one prescription. In December 1942 the Health Department also made provision for proprietary medicines to be supplied when no other form was available. However, these items would attract an 'extra for proprietary' charge. The Drug Tariff of September 1946 added a *New Zealand Formulary* to the *British Pharmacopoeia* and the *British Pharmaceutical Codex* as the basis of medicines which would be paid for from the Social Security Fund. About 20 unofficial drugs available in proprietary form were included in this (revised) Drug Tariff.

The citing of the *British Pharmacopoeia* and the *British Pharmaceutical Codex* in the Drug Tariff led in due course to the automatic adoption as pharmaceutical benefits of an increasingly wide range of drugs and medicines, including some items that were available solely under trade names, or available under trade names in particular strengths at high prices.

Faced with patient pressure, doctors became reluctant to prescribe drugs which were not a full charge on the Social Security Fund — so that inclusion of a product on the Drug Tariff became a prerequisite for the successful marketing of a product in New Zealand by drug manufacturers. Through the process of controlling admissions to the Drug Tariff, the Health

Department effectively gained control of the admission of new drugs to the New Zealand market. In theory, therefore, the Department was also in a powerful position to negotiate prices with drug manufacturers.

Large numbers of prescriptions began to be a charge on the Social Security Fund. During the first full year of operation of pharmaceutical benefits from 1942 to 1943 (year ended 31 March) 3.5 million scripts were claimed on the Social Security Fund (Health Department Annual Report, 1943). The number of prescriptions rose to 4.75 million in 1944 and to 5.5 million in 1946. The average number of scripts per pharmacy increased from 3000 per annum in 1939 to almost 10,000 in 1946 (Dodds, 1951; Joint Reviewing Committee on Prescription Pricing, 1951).

Pharmaceutical Benefits and the State

Alongside the Labour Government's commitment to promote access to health care existed another objective — to promote industry and employment in New Zealand. The Government elected in 1935 was committed to intervention in the economy. Its aims were to reduce dependence on overseas markets and to insulate New Zealand from external pressures by promoting industry at home.

The Social Security Act 1938 more or less coincided with Labour's Import Controls Regulations, designed mainly to conserve overseas funds (Baker, 1965; Hawke, 1985). This system of controls over imports was one way of controlling the use of resources and the direction of the expansion of industry. As in other industries, the importing of pharmaceutical products made or capable of being made in New Zealand was restricted. The leading advocate within the Government of an interventionist, pro-industrialization policy was the Department of Industries and Commerce (Gould, 1982: 108). This Department was concerned to see that in the interests of conserving overseas exchange funds, New Zealand manufacturing resources were 'used to the greatest extent possible in supplying the country's pharmaceutical needs' (Lewin, 1967a).

Because of the policy of encouraging local industry from the late 1930s, both New Zealand and foreign companies considering manufacturing were in a position to be offered a protected market. Overseas companies exporting finished products to New Zealand could be persuaded to complete some of the processes in New Zealand, or be denied import licences if they did not (Gould, 1982: 107). Before the Second World War many foreign companies, such as electrical machinery and chemical companies, and especially pharmaceutical and paint producers, began manufacturing in New Zealand (Deane, 1970: 124–5).

Pharmaceuticals attracted several American companies. One of the largest was Sterling Pharmaceuticals (NZ) Ltd, which began local manufacture in 1939. A New Zealand subsidiary of Merck Sharp and Dohme was established in 1962, with one quarter of the equity held in New Zealand. British investment in pharmaceuticals was not as extensive as American investment, and all the British firms, such as Glaxo Laboratories (NZ) Ltd, were 100 per cent overseas owned (Deane, 1970).

As early as 1936 Labour passed an Industrial Efficiency Act to foster new industry and to reorganize existing industry. The following year a Bureau of Industry in the Department of Industries and Commerce was given wide powers to plan new industry and to license entry into certain occupations — for example pharmacy. Pharmacy in New Zealand had been found to be in a 'parlous' condition in 1936 by a parliamentary inquiry prompted by the impending 'invasion' by the international retail chain of Boots. The Bureau of Industry set up a number of industry plans, including one for pharmacy in 1938, overseen by the Pharmacy Plan Industrial Committee (PPIC). This Committee's 'express purpose' was the economic rehabilitation of pharmacy — 'getting the small man on his feet'. At this stage Social Security dispensing appeared to the Committee as a 'burden' about to be taken on by pharmacists.

The Committee drew up its Official Schedules and Rules for Prescription Pricing in 1939, to provide a scale of charges to the public to be observed by chemists for medicines ordered on a doctor's prescription. The wholesale price of drug ingredients, as quoted by the Wholesale Druggists Federation, formed the basis upon which prescriptions were priced. To this wholesale price was added a percentage or 'loading'; to this 'loaded' price was added a further percentage, or breakage fee, calculated on the gross price, to allow for wastage when bulk items were broken up for use in small quantities in prescriptions. The pharmacist also received a fixed dispensing fee.

When pharmaceutical benefits were introduced in 1941 these same Rules were taken over by the Department of Health, to form part of the Drug Tariff. Thus pharmacists' reimbursement by the Department, written into the regulations of the first Drug Tariff in 1941, was based upon their proportion of income for dispensing in 1939:

In fixing the prices payable to contractors the ruling prices existing in 1939 [were] taken as a basis, that is to say the price prevailing when dispensing formed only 30% of turnover but returned 55% of the chemists' net profit (Department of Health, *circa* 1947).

The effective voting strength on the PPIC was in the hands of the pharmacists themselves who were represented by two Pharmacy Board nominees and one from the United Friendly Societies. The Department of Health complained about its lack of representation on the Committee, yet remained absent and therefore had no direct role in determining the basis upon which reimbursement to pharmacists should be made (Maclean, 1944).

The Committee continued to promote the chemists' cause, to control manufacturers' and wholesalers' margins and hence drug prices and reimbursement to chemists, and to revise their schedules. It soon controlled the expenditure of a great deal of Government money which by 1950 covered eight million prescriptions valued at £2 million. By the mid-1950s the Committee was superseded by the Price Tribunal which seemed to be 'trying to dictate instead of trying to understand and continue with [the Committee's] previous policy of giving formal approval to prices' (Notes of a deputation from the PPIC, 1952).

The Labour Government also took action to restrain rising prices, initially as a war-time emergency measure. In 1939 the Price Investigation Regulations, administered by the Department of Industries and Commerce, provided for control of prices by a Price Investigation Tribunal (later simply the Price Tribunal). The general effect of these regulations was that, for a wide range of goods and services, rises in prices had to be justified if challenged. Much later, wide powers were conferred on the Price Tribunal by the Control of Prices Act 1947, so that it could investigate and inquire into price applications, hear submissions and evidence from interested parties, and fix prices on goods or services (Lewin, 1967b). The Price Tribunal continued to function until replaced by the Commerce Commission, set up by the Commerce Act, 1975. But in spite of its announced intention, price control developed into a cost-plus adjustment process (Condliffe, 1959: 92).

Medicines, in common with other products, were subject to price control under the Control of Prices Act 1947, and fitted into this cost-plus adjustment pattern. Drugs were distributed by wholesalers to retail chemists, on behalf of the manufacturer or importer, which meant that both the manufacturer or importer and the wholesaler received margins of profit. Sales of drugs by importers to wholesalers were controlled by special approvals issued in the Control of Prices Act, which fixed the maximum price allowable to each importer and product. These prices were based upon the landed cost of drugs in New Zealand — rather than the prices ruling in the country of origin. Profit margins or mark-ups of importers and wholesalers were set by the Price Control Division of the Department of Industries and Commerce, taking into account 'heavy expenditure' on detailing and sales promotion.

In April 1967, J.P. Lewin acting for the Secretary of Industries and Commerce wrote to the Chairman of the Public Expenditure Committee:

> The Control of Prices Act provides adequate powers to deal with the problem [of] . . . the adverse disparity which exists in the case of some drugs between export prices and the prices at which supplies are made available on the domestic market in the country of manufacture. For some years, however, it has not been policy to apply these powers rigorously and the tendency has been to exercise price monitorship rather than stringent control (Lewin, 1967b).

The Producers

The modern drug industry can be traced to the nineteenth century wholesale druggist and patent remedy business. However, its more recent origins lie in the late 1930s chemical, and particularly dyestuffs, industry (Blum et al., 1981). More than 95 per cent of today's products were unknown in 1950. During the 1920s some six drugs, either singly or in combination, accounted for more than 60 per cent of the prescriptions written (Cooper, 1977). A very limited range of products was available in the 1930s, and natural products such as morphine, digitalis, and quinine were very important. Only a few synthetic compounds, such as aspirin and the barbiturates, were available.

The first sulphonamide antibacterial drugs were prepared in the 1930s, and the possibilities of other synthetic drugs were just beginning to be recognized. Since their synthesis was soon followed by the appearance of strains of bacteria resistant to their actions, researchers attempted to discover more effective antibacterials. The proliferation of antibiotics (antibacterials derived from 'living' material) such as penicillin, streptomycin (the first broad-spectrum antibiotic), and tetracycline was then followed by the psychoactive drugs of the 1950s. New companies, whose parent companies' major interests were in fine chemicals, oil, food, dyestuffs, and perfumery, entered the field (Thornber, 1986: 169).

From the mid-1930s the success of the international drug companies increasingly depended on the acknowledgement of their patents, that is, the international recognition of their rights to legal monopoly granted for a specified period (Lang, 1974: 23–5). A company with a patented drug could license other companies to manufacture and sell the drug, in return for the payment of royalties. Changes in patent law during the 1950s allowed the patenting not only of a new drug but also of variations on the existing one. Some countries granted patents only on the process (i.e. specific methods) for making a drug; some granted patents for the process, as well as the products. By 1960, New Zealand granted patents for both process and products for a term of 16 years (Thompson, 1962). Some countries which granted product patents made provision for compulsory licensing, enabling licensees (in New Zealand for example) to import and use the invention without further permission from the patentee, thus limiting the monopoly effects of the patent.

Granting to manufacturers exclusive rights for the sale of drugs under New Zealand patents meant that purchase by the Health Department from other lower-priced sources could be difficult and controversial. In 1962, for example, the Department of Health invited tenders for chlorpromazine hydrochloride, a tranquillizer developed in 1951 by Rhône Poulenc of Paris. For some years all purchases had been made from May and Baker (NZ) Ltd who held, through their parent English company, exclusive rights under New Zealand patents to the drug's sale in New Zealand.

May and Baker (NZ) Ltd tendered at £5,568/15/- for 45,000 100 mg tablets. A separate licence holder, Bamford Bros. of Canada, put in a price of £1,442/16/3; a third tender of £1,084/16/5 came from Jules R. Gilbert. The Bamford tender was accepted, the manufacturers being Paul Maney Laboratories, Canada, who paid royalties on the manufacture of the drug to Rhône Poulenc.

In spite of the protests of May and Baker that this purchase was an infringement of their patent rights, the Crown Law Office ruled that the Department of Health had acted properly under Section 55 of the Patents Act 1953, which allowed Government Departments to ignore a patent when purchasing drugs for use in State institutions. Less than a year later, however, the Government found that it was difficult to re-order in the light of litigation in Canada, and the continued protests of May and Baker. In 1965 the Government proceeded as it had done in 1962 (with a saving of nearly £9,000 on the order). But when the Government tried again in 1966, May and Baker sought compensation for lost income in respect of earlier purchases (Submissions to the Public Expenditure Committee by the Department of Health, 1968). In the light of difficulties of this kind the Department of Health suggested several times over the years that Government seek to reduce the life of patents from 16 years to, for example, 10 years, to avoid payment of monopoly prices for longer than necessary.

From the early 1950s production and promotion in the pharmaceutical industry focused more and more upon brand-name products. Where a drug was sold by several companies in the same country, it could be marketed under different brand names. In New Zealand the number of proprietary or brand-name products on the Drug Tariff increased steadily. By 1953 about 100 proprietary drugs were included; by 1959 the number had increased to about 1000. By 1963 about two-thirds of all prescriptions involved the supply of a proprietary or brand-name drug (Hayes, 1963).

The unique character of the prescription drug market in which patients have limited control over what they pay for, and where a third party (the State) covers the cost, encouraged the industry to concentrate on addressing the doctors — who were persuaded to remain loyal to brand names. Doctors were not only selecting a particular product by specifying a brand name, they were also selecting a particular manufacturer. In New Zealand, importers and wholesalers made little attempt to inform doctors of the prices of their wares. Their advertising often emphasized that a particular preparation was 'free under social security'.

By the mid-1960s most branded prescription medicines were imported into New Zealand, mainly as finished goods in packaged form under the trade mark of the manufacturer. However, there was still some local manufacture of these products, and also some local packaging of tablets imported in bulk. Import controls had steadily pushed into local manufacture under licence some overseas manufacturers who would have preferred to supply New Zealand's small market from factories elsewhere (Hayes, 1963). By this time most internationally known drug manufacturers had New Zealand subsidiaries or agents, usually supplied chiefly from their British or Australian factories.

At the same time, New Zealand Governments were aware that a variety of sources of supply could encourage competition and keep drug costs lower than local manufacture allowed: 'When drugs are manufactured in New Zealand, even [just] processing and packaging, production costs tend to be high' (Report of the Special Committee on Pharmaceutical Benefits, 1961–1962).

As in other countries where the State has become the main purchaser and regulator of prescription medicines, drug manufacturers, importers, and wholesalers formed trade associations in New Zealand. Agents and manufacturers of 'ethical proprietary medicines' met in Auckland in December 1941 soon after the introduction of pharmaceutical benefits the previous May. They decided to form an association to provide an official body to co-operate and consult with the Health Department on matters affecting the interests of its members.

The New Zealand Ethical Pharmaceuticals Association (later the New Zealand Pharmaceutical Manufacturers Association, now the Researched Medicines Industry) was formed in 1961. With one or two exceptions its members represented the major drug companies marketing prescription drugs in New Zealand. For some years the managing director of May and

Baker, who was also the president of the New Zealand Pharmaceutical Manufacturers Association, was the Pharmaceutical Manufacturers Association representative on the Health Department's Pharmaceutical Advisory Committee.

The Prescribers

Medical practitioners as 'gatekeepers' are still a crucial part of the pharmaceutical benefits scheme. Indeed, the family doctor, readily available to patients, is at the centre of New Zealand's medical services. But just how to provide general practitioner services, the 'medical benefits' of the 1938 Social Security Act, proved to be an almost insoluble problem for the first Labour Government. Its quarrel with the powerful New Zealand branch of the British Medical Association (BMA) simmered for three years (Hanson, 1980; Bolitho, 1984).

The Labour Government originally proposed a general contract based on an annual fee or 'capitation' for each patient registered. However, the doctors' representatives strongly opposed any scheme in which the doctor was paid by salary, capitation, or a fixed total fee for service. Such a scheme amounted to a salaried service, and in the view of the BMA was unacceptable. It seemed to threaten the doctors' professional independence, to interfere with what they considered to be the proper doctor-patient relationship, and to draw them too closely into the control of the State.

The Government was in general opposed to any scheme which sought to interpose a fee between patient and doctor. However, in August 1941 Arnold Nordmeyer, the Minister of Health, offered doctors a fee for service (Hanson, 1980). Thus a new scheme, whereby doctors agreed to afford medical benefits to patients at the cost of the Social Security Fund, was finally introduced through the Social Security (Medical Benefits) Regulations 1941. The Social Security Amendment Act 1949 allowed doctors to charge the patient an amount over and above the Social Security subsidy payment. They had successfully defended the right to charge a fee and, significantly, remained outside the comprehensive health services scheme which the Labour Government envisaged in 1938.

The introduction of pharmaceutical benefits had no immediate effect on doctors' procedures or powers, apart from limiting periods of treatment or supply under one prescription. They did not accept any official responsibility to the Health Department, and they had no obligations to the Department to prescribe in a particular manner, or to use any standard form; they remained untouched by the new order and adopted no formal position towards it. 'It is extremely difficult to exert any effective influence on the manner of doctors' prescribing and the matter can only proceed by taking appropriate action in individual cases of excessive prescribing' (Maclean, 1944). The Department of Health warned doctors repeatedly about 'wasteful and extravagant prescribing'.

As well as determining which medicine the patient would have, doctors were well represented on the Pharmacology and Therapeutics Advisory Committee which, since the early 1950s, has considered additions and amendments to the Drug Tariff. The Committee was enlarged in 1957 to make it more widely representative of doctors and to increase its efficiency. Chaired by the Director of the Clinical Services Division of the Department of Health, it now included two physicians, a surgeon, a paediatrician, a senior member of the staff of the Medical School, and a general practitioner (who was also a qualified pharmacist) (Report of the Special Committee on Pharmaceutical Benefits 1961-1962; 1963).

From the later 1950s the Department of Health began to maintain contact with doctors through three series of printed circular letters — *Notes on Prescription Costs*, *Therapeutic Notes* (short articles on drug treatment), and the *Clinical Services Letters*, which supplied information about changes in the Drug Tariff and rises in the cost of pharmaceutical benefits. The Department, however, has been able to exert little influence over the doctors' freedom to prescribe.

The Dispensers

In 1941 pharmacists were still the key 'makers' of medicines, which they compounded in final form. This role, together with the dramatic increase in dispensing, ensured pharmacists' prosperity in New Zealand for the next 50 years. Their professional role, however, began to decline, especially from the early 1950s when manufacturers began to distribute medicines which were almost entirely compounded in various dosages under a brand name (James, 1981).

When pharmaceutical benefits were introduced in 1941, the Chemists' Service Guild, acting on behalf of more than 500 pharmacists all over New Zealand, signed up formal contracts with the Minister of Health. The 'surprisingly favourable terms' had been negotiated by the Director of Pharmacy of the PPIC, which represented chemists' interests inside the Government. Indeed, the Committee's central interest seemed to be to support and promote pharmacy rather than 'health'. Some pharmacists began to advertise 'Free Doctor's Medicine' which was considered undignified by the Committee.

As well as supervising the licensing of all pharmacies, the Committee had been given the right to fix drug prices. The Committee issued its Rules for Prescription Pricing in 1939, setting out the sums to be charged for medicines, either when included in prescriptions, or when sold or dispensed uncompounded.

These same rules for prescription pricing formed part of the regulations of the first Drug Tariff in 1941, and became the basis of Government reimbursement to chemists for the next 50 years. Thus, instead of setting prices to individual customers or the general public, as was the original intention, the Committee's prices now applied to one main customer, the Government itself.

The pricing system was based principally on a percentage of the current wholesale prices of dispensary requirements. Pharmacists were paid the wholesale price of the ingredients plus a mark-up of 50 per cent on most items (sometimes more), a standard container charge, a dispensing fee, a breakage fee and finally an overall discount of 2.5 per cent. (Reimbursement to pharmacists is still based on all these calculations apart from the breakage fee — the dispensing fee has increased; the mark-up on the wholesale price of ingredients has decreased.) Reimbursement by the Health Department was not necessarily tied to the actual cost to the pharmacist, who was still free to buy at discounted prices from wholesalers.

After re-labelling and re-packing in appropriate quantities, most proprietary lines, which carried a mark-up of 50 per cent, were dispensed by the pharmacist in the same form as they were received. C.H. Farquarson, the Health Department's pharmacist, pointed out in a 1947 memorandum to the Director of the Division of Health Benefits that there was no doubt that 'the incidence of these high priced prescriptions is increasing and the position has materially altered since the percentage on these lines was first determined in 1941' (Farquarson, 1947). The higher the manufacturer's price (and consequently the wholesale price), the higher the profits were to chemists as well as to wholesalers.

Dispensing quickly became the most significant proportion of a pharmacist's income. For the year ended 31 March 1944, the Social Security Fund paid to contracting chemists the sum of over £700,000, nearly 65 per cent of their total turnover of 1939 (Department of Health, *circa* 1947). As early as 1944 when dispensing had almost doubled, a pharmacist was receiving, perhaps from that source alone, an income approximately equal to his total income *from all sources* in 1939. Almost all of the pharmacist's income from dispensing was now paid by the Health Department from the Social Security Fund and later from Vote: Health.

From the mid-1950s to the late 1980s, the proportion of pharmacists' income derived from Social Security dispensing steadily increased. For example, income from Social Security dispensing, as a percentage of income from all sources, was 34 per cent in 1955, 42 per cent in 1960, 44 per cent in 1966 and in 1971, 46 per cent in 1976, 49 per cent in 1981, 56 per cent in 1986, and 57 per cent in 1988 (Hayes, 1963; Pharmacy Guild of New Zealand, 1989).

Pharmacists' incomes 'are greater than is justifiable' complained the Deputy Director-General of Health, Dr Duncan Cook, to D.G. Sullivan, Minister of Industries and Commerce, as early as April 1945. 'In other words . . . the State is being mulcted in much too high a sum for the services rendered. Two or three government nominees should be retired from the PPIC', he suggested, 'in favour of appointments which would be made from the Health Department' (Cook, 1945).

Every price increase gave the pharmacist an automatic increase in profit. In fact:

It pays him to buy at the dearest price and in the largest quantities in order to reap the benefit of the loading. He is in much the same position as the building contractor who completes a job on a '10 per cent on cost' basis and who finds it profitable to pay his workmen high wages and equip his building with expensive hardware. A further disquieting feature is that the retail trade no longer acts as a 'brake' on wholesale prices. The contractor has no longer an interest in obtaining supplies on the most economical terms. His customer, the Fund, is committed to payment and the higher the price the higher the profit to both wholesaler and retailer (Department of Health, *circa* 1947).

Controlling Costs

Anxiety about the rising costs of pharmaceutical benefits as a proportion of all health benefit costs surfaced at once. In March 1944 Arnold Nordmeyer as Minister of Health wished to know whether increased expenditure was due to unnecessary prescribing, prescribing of expensive drugs, increased cost of drugs due to war circumstances, introduction of expensive preparations (mainly proprietary) in recent additions to the Drug Tariff, or over-payment of pharmacists 'having regard to the value of the services rendered' (Nordmeyer, 1944). Later that same year Nordmeyer suggested to the Minister of Industries and Commerce, 'that all that was desired was to ensure that the Department responsible for payment of the prices set for prescriptions should have some more effective voice in the determination of such prices' (Nordmeyer, 1944).

That same year the Director-General of Health emphasized the method of payment to the pharmacists as the greatest cause of increasing costs of pharmaceutical benefits:

> There has necessarily been an overall increase in the wholesale cost of drugs during the war but the method of computing the amounts paid to [pharmacists] allows for profits on a percentage basis and as the wholesale costs rise so do actual profits made by the contractors. The method of calculating the retail cost of drugs appears to be too inflexible and while a 50% profit may be reasonable on an article costing a few pence, it appears to be grossly excessive when applied to high priced drugs. To take an extreme case a prescription was issued for a certain Hormone preparation which is already packed for issue to the patient and cost the Fund £14.7.5. Of this the contractor's profit was £4.16.0 and no dispensing procedure was required (Maclean, 1944).

Four years later, in 1948, the Medical Services Committee (Cleary Committee) reported that the chief factors which had caused the heavy annual increase in the cost of these benefits to the Social Security Fund were unnecessary prescribing and the use of new and expensive drugs. The Committee accordingly recommended to Government that, except for specific cases such as the supply of insulin to a diabetic, the principle of part payment by the patient be adopted. However, its advice was not followed.

The cost of drug content was examined. For example, the ingredient content of prescriptions rose from $0.32 million in 1943 to $13.6 million in 1966. Yet the number of prescriptions increased from 3.5 to 17 million in this time (Public Expenditure Committee, 1968). In other words, the cost of drug content rose about 40 times while the number of prescriptions dispensed rose five-fold. T.L. Hayes, the Assistant Director of Clinical Services Division, pointed out that: 'The effect of the rise in ingredient cost is accentuated because of the on-cost allowed to chemists' (Hayes, 1963).

By 1960 pharmaceutical benefits were the single most expensive health benefit. By the mid-1960s the cost of pharmaceutical benefits was rising at the rate of 10 per cent per annum, that is by about $2 million a year at a time of low inflation. Much of this sum had to be paid overseas, thus aggravating New Zealand's balance of payments difficulties. The Department of Industries and Commerce was 'concerned to see that in the interests of conserving overseas exchange funds New Zealand manufacturing resources are used to the greatest extent possible in supplying the country's pharmaceutical needs' (Lewin, 1967a).

In 1966 Dr Hayes, Assistant Director of the Clinical Services Division, concluded that:

All known methods of control have been used over recent years in an endeavour to curb this rising cost, but with little success. The number of prescriptions per head has now stabilised, and there is now little doubt that the chief reason for the increase in cost . . . is due to the high cost of *new* drugs. Two important questions remain unanswered (a) Are we paying too much for the newly discovered drugs? (b) Are we buying our drugs on the best market, and by the best methods? (Hayes, 1966).

Government committees of inquiry continued to study pharmaceutical benefits and the problems of their provision. 'There is some apprehension that prices for drugs are being loaded against NZ', the Director-General of Health wrote to the Minister of Finance in 1966 when raising with Treasury the idea of another investigation on pharmaceutical costs. The Public Expenditure Committee began sitting in March 1967, and received submissions from a wide range of groups. The Committee's report recommended among other things that the State establish a medicines commission (whose functions were spelt out in detail), that the period of patent protection for drugs be reduced from 16 years to 10 years, and that the payment to retail pharmacists be reapportioned. Most of these recommendations, however, languished after detailed investigations.

Some pharmacists commented on the Government's failure to establish proper structures and procedures. One commented (anonymously):

Generally speaking, I feel the [Health] Department's main concern is that of containing the costs within certain limits which are influenced by Treasury, and not by health considerations. This leads to a policy which seems to be without any definite basis. The policy is just 'ad hoc' decisions thought up to contain costs (Report of the Special Committee on Pharmaceutical Benefits 1961–1962; 1963).

A flat-rate charge for prescriptions was proposed many times over the years in response to mounting anxiety over the cost of pharmaceutical benefits. Certain medicines were subject to a part charge, some could be obtained only through public hospital dispensaries, and some required endorsements of various kinds. However, a flat-rate charge for prescriptions was a difficult political decision for Governments of either major political party, and such a charge was not levied for the first time until 1985.

Conclusion

Pharmaceutical benefits have been a key element of the Welfare State in New Zealand since 1941. Yet prescription medicines are also the source of profits of an enormously powerful international industry. Since 1941, the Health Department has been committed to providing safe and effective medicines at no or low cost to the patient. This policy has posed several dilemmas — all of which are linked to the State's intractable problem of controlling the costs of pharmaceutical benefits.

No Government could have foreseen the huge cost of providing the products from a rapidly growing industry which, in its modern form, was almost non-existent in the 1940s. The spectacular growth, especially to the 1960s, of brand-name or proprietary medicines, has resulted in an increase in the cost of pharmaceutical benefits. However, no Government could withhold these products once they became available.

At the same time, the Department of Health has had at its disposal two effective means by which it could have greater leverage in its bargaining with drug companies: prices, and entry to the market. However, until recently the Health Department has seen its role as an efficient administrator of the pharmaceutical benefit scheme, rather than as a tough commercial negotiator with the industry.

Technically this price 'control' of pharmaceuticals, that is the setting of maximum prices to importers and wholesalers, was the function of the former Department of Trade and Industry until 1986. The State has not had a single policy on pharmaceutical benefits. The State as funder, through the Department of Health's 'health' or free medicines policy, has co-existed with the State as economic manager, through the 'industry' policy of the former Department of Trade and Industry.

This chapter has examined the intricate machinery set in place for the provision of pharmaceutical benefits from the time of their introduction in 1941. The volume and cost of prescriptions has risen enormously since that time, and recently some important changes have been made in the way in which pharmaceutical benefits are provided. However, the interests of those who produce, fund, prescribe, and dispense medicines remain fundamentally unchanged 50 years later — as does the problem for Government of controlling the costs of pharmaceutical benefits.

References

Background notes on the recommendations made by the Public Expenditure Committee (5 March 1969) Health Series 1, Sub-series 208/53/1 file 35927.

Baker, J.V.T. (1965) *Official History of New Zealand in the Second World War 1939-45: the New Zealand People at War. War Economy* Wellington, Historical Publications Branch, Department of Internal Affairs.

Blum, R. (1981) Introduction. In Blum, R. et al. (ed) *Pharmaceuticals and Health Policy: International Perspectives on Provision and Control of Medicines* 11-26, London, Croom Helm.

Bolitho, D.G. (1984) Some financial and medico-political aspects of the New Zealand medical profession's reaction to the introduction of Social Security *New Zealand Journal of History* 18: 34–49.

Condliffe, J.B. (1959) *The Welfare State in New Zealand* London, Allen and Unwin.

Cook, D. (11 April 1945) Memorandum to D.G. Sullivan. Health Series 1, Sub-series 208 file 17977.

Cooper, M.H. (1977) Substitute competition and the international pharmaceutical industry *Australian Journal of Pharmaceutical Sciences* 6: 113–18.

Deane, R.S. (1970) *Foreign Investment in New Zealand Manufacturing* Wellington, Sweet and Maxwell.

Department of Health Annual Report (1943) Appendices to the Journals of the House of Representatives A to L, H31.

Department of Health (*circa* 1947) Memorandum on Pharmaceutical Costs. Health Series 1, Sub-series 208 file 17977.

Dodds, D.S. (31 March 1951) Notes for the review of dispensing fees. Industries and Commerce Series 4, 22/1/1 (accession 3852).

Farquarson, D.H. (1947) Memorandum for Director of Division of Health Benefits. Health Series 1 file 5467.

Gould, J.D. (1982) *The Rake's Progress? The New Zealand Economy since 1945* Auckland, Hodder and Stoughton.

Hanson, E. (1980) *The Politics of Social Security: The 1938 Act and some Later Developments* Auckland, Auckland University Press/Oxford University Press.

Hawke, G.R. (1985) *The Making of New Zealand: An Economic History* Cambridge, Cambridge University Press.

Hayes, T.L. (27 March 1963) New Zealand Report on the market for pharmaceuticals Health Series 1, Sub-series 208 file 33232.

Hayes, T.L. (11 February 1966) Note on overseas tour of duty, Health Series 1, Sub-series 208 file 33232.

James, B. (1981) *The Marketing of Generic Drugs* London, Associated Business Press.

Joint Reviewing Committee on Prescription Pricing (May 1951). Inquiry on Prescription Pricing. Health Series 1, Sub-series 208 file 25408.

Lang, R.W. (1974) *The Politics of Drugs: A Comparative Pressure-Group Study of the Canadian Pharmaceutical Manufacturers Association and the Association of the British Pharmaceutical Industry, 1930–1970* Farnborough, Hants, Saxon House.

Lewin, J.P. (6 March 1967a) for Secretary, Department of Industries and Commerce. Memorandum to the Chairman of the Public Expenditure Committee. Legislative Department Series 1/1968/7/1.

Lewin, J.P. (19 April 1967b) for Secretary, Department of Industries and Commerce. Memorandum to the Chairman of the Public Expenditure Committee: Inquiry into the cost of drugs in relation to pharmaceutical benefits. Legislative Department Series 1/1968/7/1.

Maclean, F.S., Director-General of Health (14 July 1944): Memorandum on Pharmaceutical Supplies Benefits Health Series 1, Sub-series 208 file 17977.

Nordmeyer, A. (26 September 1944) to Sullivan, D.G., Minister of Industries and Commerce. Health Series 1, Sub-series 208 file 17977.

Notes of a deputation from the Pharmacy Plan Industrial Committee (PPIC) to J.T. Watts, Minister of Industries and Commerce (14 May 1952) Basis of pricing and future of the Pharmacy Plan Industrial Committee Health Series 1 Sub-series 208 file 25408.

Public Expenditure Committee (1968) Draft Report Legislative Department Series 1/1968/7/1 file 80.

Report of the Special Committee on Pharmaceutical Benefits 1961-1962 (March 1963) Wellington, Government Printer.

Submissions to the National Health and Superannuation Committee (1938). Legislative Department Series 1/1938/10 Box 897, Section 2 A 19-26.

Submissions to the Public Expenditure Committee by the Department of Health (1968). Review and general prospects Appendix G. Legislative Department Series 1/1968/7/1.

Thompson, A.W.S. (7 August 1962) Proposed abolition of patents on drugs in New Zealand. Health Series 1, Sub-series 208 file 33232.

Thornber, C.W. (1986) The Pharmaceutical Industry. In Heaton, C.A. (ed) *The Chemical Industry* 169-228 London, Blackie.

The Changing Role of Pharmacists and the Distribution of Pharmaceuticals in New Zealand

Pauline Norris

This chapter is about the changing role of pharmacy in New Zealand. Pharmacists work within a complex network of professional, commercial, and bureaucratic relationships. Analyzing the changing nature of that system of relationships tells us much about the current arrangements governing the distribution of pharmaceuticals in New Zealand and how those arrangements can produce unexpected, and sometimes undesirable, outcomes. This is further exemplified in a close analysis of two recent policy initiatives — patient charges and substitution rights. These two case studies show how complex policy-making in the area can be and illustrate the difficulties and obstacles in the way of securing key policy objectives.

From Compounder to Dispenser

In the early days New Zealand pharmacists manufactured most or all of the medicines they dispensed. This required considerable manual skills and dexterity as well as knowledge of drug formulation. These were learnt through an apprenticeship system (Combes, 1981). Pharmacists also played an important role as advisers and were a major source of care for many people for whom doctors were unavailable or too expensive.

Since early this century large companies have become increasingly involved in the mass manufacture of drugs. The 1909 discovery of salvarsan, an arsenic compound active against syphilis, heralded the beginning of the 'drug revolution'. In the 1930s and 1940s the discovery and synthesis of new drugs, particularly antibiotics, began to change the role of community pharmacists. Economies of scale in research, manufacturing, and marketing of the new drugs meant that pharmacists lost much of their small-scale manufacturing and began increasingly to deal with mass-produced drugs. The introduction of Government subsidized general practitioner visits and pharmaceuticals in 1941 increased the numbers of prescriptions dispensed, but may also have eroded the demand for health advice from pharmacists as more people could afford to seek medical attention (Combes, 1981).

The number of prescriptions dispensed continued to rise over the years (see Astrid Baker's chapter), but there was less work involved in each one. Dispensing mass-produced drugs became more common than compounding — which in the early 1990s is quite infrequent. Pharmacy technology changed: prescriptions which were originally recorded in books, then on microfilm, are now mostly stored on computer. Pharmacy education also changed. Apprenticeships gradually gave way to technical college courses, and now, in the early 1990s, all pharmacy education is to be located at Otago University.

From 'Physician's Chef' to Drug Expert?

Pharmacists lost their traditional manufacturing role, but new technology also created new opportunities for them. Both overseas and in New Zealand many pharmacists have argued that the development of modern drug therapy has created the need for a 'drug expert' to advise patients and doctors. They argue that pharmacists are the only people trained specifically in the actions and effects of the new and more potent drugs, and are in a position to ensure that patients get the appropriate drugs in the right dosage.

This new movement, referred to as 'clinical pharmacy', developed first in the United States (US), but is becoming increasingly important in other countries, including Britain and New Zealand.

In the US the change from 'physician's chef' to drug educator has been quite dramatic. In a 1966 pharmacy textbook budding pharmacists were advised that 'the pharmacist is not authorised to discuss with the patron either the nature or the use of the drug. Questions on these matters are courteously referred to the physician' (Deno et al., 1966: 69). By 1989 Boards of Pharmacy in 10 States of the US required pharmacists to keep patient medication profiles, and in 13, pharmacists were required to give some information about drugs to customers. Six States required both (Campbell et al., 1989).

In the late 1960s and 1970s concern in the US over the incidence of adverse effects of drugs gave pharmacists the opportunity to argue that they had an important role in ensuring appropriate drug therapy. It has been suggested that the introduction of new dispensing technology and the use of technicians to do routine dispensing tasks in the hospital setting allowed pharmacists to develop these new aspects of their role (Broadhead and Facchinetti, 1985).

In the United Kingdom (UK) debate over the role of pharmacists led to the Pharmaceutical Society supporting an independent inquiry into pharmacy (Holloway, Jewson, and Mason, 1986). The resulting report, referred to as 'The Nuffield Report', came out strongly in favour of a more clinical role for pharmacists. It concluded:

Dispensing will continue to be an important activity within pharmacies but the pharmacist's future professional role should be seen in terms of greater collaboration with other health care professionals, particularly GPs; and greater involvement with members of the public (Committee of Inquiry of the Nuffield Foundation, 1986).

Pharmacy journals are reflecting this change in emphasis. In 1982 the American Society of Hospital Pharmacists started a journal called *Clinical Pharmacy*. It contains reports of research, articles on how to recognize and treat various diseases, and a considerable amount of drug company advertising. This suggests that the pharmaceutical industry believes pharmacists are exerting increasing influence on choices about which drugs are used. More general pharmacy journals in the US and the UK (for example *The Pharmaceutical Journal*) now include sections describing, for example, how to identify and intervene in illnesses such as eating disorders, or what drug treatments are recommended for problems such as anxiety, along with discussions of the contribution that pharmacists could make towards improving health care.

An important claim of clinical pharmacy is that pharmacists' knowledge of drugs and their effects allows them to encourage more cost-effective prescribing. In the UK some pharmacists are employed in health centres to research the safety, efficacy, and cost of new drugs, prepare up-to-date formularies, and monitor the doctors' prescribing (Brompton, 1988).

Encouraging rational and cost-effective prescribing has also long been a role for one or two New Zealand pharmacists. The Department of Health employs a Visiting Pharmacist to visit general practitioners. Part of his role is to discuss prescribing with the doctors he visits, and to suggest non-pharmaceutical ways of dealing with minor conditions. He sometimes takes a bundle of the doctor's prescriptions with him to use as a focus for discussion. The potential of this role for pharmacists has been explored more fully in the Nelson General Practice Prescribing Project (see Helen Clark's chapter).

In the US and the UK the biggest changes toward clinical pharmacy have occurred in the hospital setting (Holloway, Jewson, and Mason, 1986).

In New Zealand, pharmacists are involved in clinical activities in many hospitals. For example, at Dunedin Hospital pharmacists are involved in ward meetings, patient discussion groups, and in medication counselling when patients are discharged. They regularly publish a bulletin which includes information about the hospital formulary, new products, cost-saving measures, and clinical information. This is distributed to medical staff in the hospital and to general practices and pharmacies in Dunedin (Cammell, 1990).

A recent survey of New Zealand retail pharmacists found that many supported a greater role in patient counselling, drug selection, monitoring and advising on adverse reactions, and diagnosing and treating minor conditions (Roseman, 1989). However, as outlined later in this chapter, retail pharmacists' income is still determined by the number of prescriptions they dispense, the value of the drugs dispensed, and retail sales.

Thus there are two views of what pharmacists do. One is of a distributor of medicines who takes 'pills from a larger bottle and puts them in a smaller one' (United Nations Conference on Trade and Development, 1981) while another sees pharmacists as primary health care providers with an important contribution to make towards improving the use of medicines. Many pharmacists appear to support the latter view, while their remuneration is based more on the former. Each view has different implications for policy, particularly how pharmacists should be paid. If we see pharmacists as primary health care providers, perhaps they should be paid for this. The Nuffield Report recommended reducing the payments for prescriptions dispensed and creating separate payments for other professional activities, perhaps including domiciliary activities, health education work done with doctors to improve prescribing, and advice to patients. At present in New Zealand the financial incentives for pharmacists often work against the development of a more clinical role.

Pharmacy in New Zealand

In New Zealand most prescription drugs are distributed through retail pharmacies. Most New Zealand pharmacists work in the retail sector. Only about 300 out of the approximately 2500 practising pharmacists work in the hospital sector (New Zealand Hospital Pharmacists Association, 1990). A very small number work in other areas such as industrial pharmacy. In 1980 it was estimated that only about 14 to 17 per cent of pharmaceuticals were distributed through the hospital sector. Hospitals mostly buy direct from manufacturers or importers (Newell et al., 1980).

The way pharmaceuticals are distributed through retail pharmacies contributes to medicines expenditure both directly and indirectly. Newell et al. (1980) calculated that in 1979, 32 cents of every dollar of medicines expenditure went to retail pharmacies, and 10 cents went to wholesalers. Although this was slightly below the average of 14 other developed countries, and the figures are probably lower now, this shows that the distribution network is an important factor in the cost of medicines. In this chapter I argue that the way pharmaceutical distribution is organized can also affect the quantity and kinds of medicines New Zealand buys.

New Zealand retail pharmacies are privately owned, and contract with the Health Department. There are about 1100 around the country (Pharmaceutical Society, 1988), in which over 2000 pharmacists work (Department of Statistics, 1990). Pharmacies can only be owned by registered pharmacists. This is not the case in some other countries. At the time of writing pharmacies must be owned by pharmacists (either individually or in partnerships) or by companies in which pharmacists own at least 75 per cent of the share capital. No person or company may own more than one pharmacy. There is a grandfather clause which allows Boots and the United Friendly Society pharmacies to operate (as they did in 1938) but their expansion and movement are restricted (Department of Trade and Industry et al., 1985). The Pharmacy Authority (a barrister or solicitor appointed by the Government) can allow departures from the usual ownership arrangements. Because of these ownership restrictions, pharmacies in New Zealand have always been small, usually involving one or two registered pharmacists. During the 1980s there has been pressure to deregulate the ownership of pharmacies — resulting in a Bill introduced in 1989 which would have reduced the share pharmacists needed to own to 51 per cent (Department of Trade and Industry et al., 1985; Pharmacy Bill, 1989).

Pharmacists contract with the Department of Health to dispense medicines covered by the Pharmaceutical Benefits Scheme. Just over half of the average pharmacy's income comes from this dispensing (Pharmacy Guild of New Zealand, 1989). The other half comes from the sale of over-the-counter medicines, related pharmaceutical goods, cosmetics, and so on. Pharmacies vary greatly in how much they rely on retail sales for their income. Mainstreet pharmacies in cities often have little dispensing business, and concentrate mainly on retailing. Suburban pharmacies traditionally rely much more on dispensing prescriptions. It has been estimated that the percentage of income from dispensing can vary from about 10 per cent in a central city location to 70 per cent in the suburbs.

Professional and Economic Relationships

In New Zealand, as elsewhere, pharmaceuticals are distributed through a complex set of relationships. There are six major actors: the manufacturer, the wholesaler, the Health Department, the doctor, the pharmacist, and the patient. The relationships between them are illustrated in the diagram below (see figure 1). Manufacturers supply drugs to wholesalers, who supply them to pharmacists. Pharmacists have shares in wholesaling companies. Prices for pharmaceuticals are negotiated between manufacturers and the Department of Health. However, the Department does not pay manufacturers directly; instead it reimburses pharmacists for part of the cost of the drugs they dispense. The rest is paid by the patient.

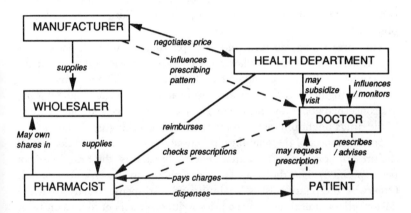

Figure 1: Model of professional relationships in the distribution of pharmaceuticals.

Unlike other industries which advertise directly to customers, pharmaceutical manufacturers intensively market their products to the doctors who prescribe them. Doctors act as gatekeepers, restricting public access to prescription-only drugs. Patients must first see a doctor, who decides what drug(s) they will receive. This is largely done in the absence of information about the real cost of the drugs. Doctors' prescribing drives the system of drug distribution. It determines what drugs are dispensed. Doctors give prescription forms to patients, who take them to pharmacists. Pharmacists and doctors do not necessarily have much contact with each other, although pharmacists are responsible for checking doctors' prescriptions.

As Ichiro Kawachi and Joel Lexchin point out in their chapter, prescriptions are written in over six out of 10 visits to the doctor. Prescriptions may be written for a particular brand of drug(s) or for the more general, generic name for each drug. The latter means that the active ingredient, formulation, and dose of the drug are given, but the actual manufacturer of the drug is not. A parallel in the non-drug field might be something like 'iodized salt' instead of 'Skellerup' brand iodized salt. All iodized salt has the same active ingredients (salt and iodine), but is sold by different companies.

Doctors are not legally allowed to direct patients to a pharmacy; they are free to choose any pharmacy. Nevertheless, there is a growing trend toward building pharmacies associated with medical centres. When patients attend a doctor's surgery which has a pharmacy attached, it is often more convenient for them to take their prescription there, but they are not legally obliged to.

The prescribing doctor, and the pharmacist who dispenses the prescription, do not usually have any contact with each other about the prescription unless this is initiated by the pharmacist. The pharmacist will probably ring the doctor if they are unsure about the doctor's handwriting, are concerned about an unusual dose, or are concerned that one or more of the drugs might interact in some way. This checking-up function is an important aspect of the relationship between pharmacists and doctors. If the prescription has been written generically (that is, the brand has not been specified) then the pharmacist can choose to dispense any brand. Otherwise the pharmacist is legally bound to dispense exactly what the doctor orders, unless they have a prior written agreement to the contrary.

The Distribution of Prescription Medicines

The distribution chain starts with manufacturers of prescription drugs. They negotiate prices for each drug with the Department of Health. They then sell the drugs to wholesalers at this price. Pharmacists buy them from pharmaceutical wholesalers. (There are exceptions, and buying direct from manufacturers seems to be becoming more common.) The pharmacist pays the price of the drugs (previously agreed between the manufacturers and the Health Department) plus the wholesale mark-up (also set by the Health Department). The wholesale mark-up is supposed to allow wholesalers to cover their costs and make some profit. Wholesalers sometimes offer discounts to pharmacists.

Pharmacists keep stocks of medicines so that normally they can quickly dispense whatever medicines are prescribed. Sometimes they need to 'borrow' them from another pharmacy or to order them. The pharmacist has usually already paid for the drugs when the patient arrives with a prescription. Therefore it is important for pharmacists to accurately estimate the kinds of drugs which will be prescribed and in what quantities. Some pharmacists say this estimation is easier in a very small town, or a suburban area where all the prescriptions come from a regular group of doctors and patients. The kinds and quantities of drugs tend to be fairly stable in those conditions so pharmacists are likely to dispense more or less the same range of drugs from month to month. Many pharmacies are now equipped with computers with stock control programs.

The set of arrangements for distributing and paying for drugs has a long history behind it, growing out of early negotiations between the Government, pharmacists, and the industry that preceded the introduction of the Pharmaceutical Benefits Scheme. These are described in Astrid Baker's chapter in this book. The result is a complex system that may have suited the economic interests of the parties at the time but cannot now be relied upon to serve the public interest in an efficient and equitable manner, frequently producing unexpected and perverse outcomes.

When prescriptions are dispensed pharmacists are responsible for collecting the patient contribution (in June 1991 $15 or $5 depending on the patient's age, employment status etc.) plus any part-charges (see below). The prescription forms are kept and at regular intervals pharmacists send bundles of these to a Department of Health prescription pricing office. This allows pharmacists to be paid for the drugs and the work involved in dispensing them. They are reimbursed for what they should have paid for the drug (that is, the price of the drug set by negotiations between the manufacturers and the Department of Health plus the wholesale mark-up). If pharmacists have bought the drugs cheaply because of wholesalers' discounts, they are still reimbursed at the same level. This means that any discounts are not passed on to the Government. (Recently, however, the margin has been adjusted to take account of the average level of discounting.) As payment for their role in drug distribution pharmacists are paid a mark-up (currently 11.28 per cent), a dispensing fee (which varies with the difficulty of dispensing), and a container fee. The patient contribution (the $5 or $15 already collected from the patient) is subtracted.

The mark-up compensates pharmacists for purchasing and storing the drugs. The dispensing fee compensates them for the work involved in dispensing the drugs. Both the dispensing fee and the mark-up are negotiated between the Department of Health and the Pharmacy Guild (which represents proprietors of retail pharmacies). Recently the Guild, and pharmacists in general, have been unhappy with the level of the mark-up and the fee, and there has been a good deal of acrimony between the two parties to the negotiation.

The Department of Health and the drug manufacturers negotiate the prices the Department will pay for each drug. In some cases the Department is not willing to pay the price wanted by the manufacturer and may decide to pay only some of the price — leaving the patient to pay the rest of the cost. This is usually in instances where an acceptable alternative is available at a lower cost. This is known as a part-charge and reflects the difference between the higher- and lower-priced products. This ability to place part-charges on drugs is an important bargaining tool for the Department. Manufacturers wish to avoid having part-charges on their drugs whenever possible, because they are obviously unpopular with patients. Doctors and pharmacists (sometimes prompted by patients) will often choose a substitute for these drugs to avoid the part-charge. This forces manufacturers to trade off the market share and the price they get for their drugs (Scott et al., 1986: 66). It also reduces the incentive for pharmacists to dispense more expensive brands of a drug.

Consequences of this Payment System

How health-care providers are paid for their work has important consequences. Remunerating pharmacists by a margin on top of the cost of the drugs they dispense has meant that as the base price of drugs has risen over the years, so has the value of the margins. In order to compensate for this the Department of Health has cut back the margins from the 50 per cent paid at the introduction of the Pharmaceutical Benefits Scheme in the early 1940s.

This 'cost-plus' method of payment means that unless the Department of Health alters the margins, pharmacists' incomes are tied to the total drug bill. That is, if the cost of drugs rises, so do pharmacists' incomes; as the total cost decreases, so do pharmacists' incomes. It also means that whenever pharmacists have an influence over what drug is prescribed or a choice of what to dispense, they have a financial incentive to choose a more expensive drug, or one which has been heavily discounted by the manufacturer or wholesaler. This kind of funding therefore produces problems for the Health Department in trying to cut the cost of drugs to the country. But it also produces problems for pharmacists. It is difficult for them to argue that their decisions are not influenced by the cost of drugs, and they are torn between maintaining their incomes and being involved in attempts to reduce the cost of drugs.

Case Study I: The Introduction and Increase in Patient Co-payments

Some important aspects of the set of relationships between general practitioners, pharmacists, patients, and the Department of Health were brought into sharp relief when a change was introduced into the system of payment for pharmaceuticals. The interdependence of the different groups and the effects of various incentives in the payment system were shown sharply by the introduction, and the subsequent increase, in the patient contribution. The example also illustrates the conflicts between the two models of pharmacy discussed above. A vision of pharmacists as medicines distributors leads to an emphasis on simply cutting costs at each part of the distribution chain, whereas a view of pharmacists as primary health care providers with an important role in monitoring and improving drug use might lead one to take into account frequency and continuity of pharmaceutical care.

In September 1988 the previous $1 flat charge for prescription items was replaced with a charge of $5, with a reduced rate of $2 for beneficiaries, children, and some other groups. Individuals were exempt from this charge after 25 items, and families after 40 items, in a calendar year (Department of Statistics, 1990). In February 1991 the National Government increased these charges and dropped the number of prescription items needed to qualify for an exemption. Adults now have to pay $15 per item, while children, superannuitants, those over the age of 65 years, beneficiaries, full-time students, and those classified as chronically ill, are charged $5 per item. Only 10 items are needed for an individual to qualify for an exemption and 15 are required for a family (Department of Health, 1991).

Table: Patient charges for pharmaceuticals

	September 1988–January 1991	February 1991–January 1992
Full rate:	$5	$15
Beneficiaries, children, chronically ill etc.:	$2	$5
Exemption after:	25 items, or 40 for a family	10 items, or 15 for a family

Further changes to the level of the patient contribution were announced in the July 1991 Budget. The National Government has proposed a new three-tier system of charges. From February 1992 'low income' people are to be charged $5 per prescription item, 'middle income' people $7.50, while 'upper income' people are to pay $20. Families are to be exempt after 15 items per year (Ashton, 1991).

The 1988 introduction of the patient charge, and the 1991 increase, led to many unexpected consequences. Firstly, and perhaps most importantly, the $2 and $5 charges seem to have led to an increase in prescriptions for longer periods.

In order to understand why, it is helpful to compare the cost to the patient of two different ways of prescribing, say a three-month supply of a drug. When a doctor prescribed a one-month supply plus two repeats the patient had to pay the patient contribution each month (for example: $5 × 3 = $15). However, there was another option. Doctors could order the whole supply to be dispensed at once. This was referred to as a 'stat' prescription (from the Latin word 'statim' meaning 'immediately'). In this case the patient could have obtained the whole three-month supply at once and paid the contribution only once (for example: $5). It is very difficult to work out how many prescriptions were being written in this way, but as well as the anecdotal accounts of doctors, pharmacists, and patients, we have evidence that since the introduction of the patient contribution the total number of dispensings has dropped. Each year numbers of prescriptions dispensed peak in the winter and drop in the summer. As well as the summer troughs, the winter peaks (that is, for the quarter ending September) per capita dropped from about 2.4 in 1988 when the $2 and $5 charges were introduced, to a low of about 1.9 in 1990 (Scott and Gini, 1991).

The effect of stat prescribing on pharmacists was quite different. Their income was adversely affected in several ways. Each time pharmacists dispensed a medicine they were paid a dispensing fee. When a patient came in with a prescription with two repeats, this represented three dispensing fees. A stat prescription represented one dispensing fee. This compounded the effects of the reduction in the wholesale margin.

Stat prescribing also made it harder for pharmacists to estimate the amount of stock they needed to keep. They needed to be able to anticipate demands for three-month supplies, and this meant they tended to need more stock, thus having more capital tied up in stock. Some pharmacists have suggested that their retail sales were also adversely affected because patients came into the pharmacy less often. As a consequence, the local general practitioner joined the Health Department as the perceived cause of pharmacy's financial problems. Some pharmacists saw doctors as protecting their own income at the expense of pharmacists.

Pharmacists argued that stat prescribing made it more difficult for them to monitor compliance with drug therapy. When patients came back for their repeats pharmacists had an opportunity to check that they were taking the medication regularly and patients had an opportunity to discuss any problems they might be having with the medication. Pharmacists also suggested that stat prescribing led to significant wastage. They told stories of people being prescribed a three-month supply of a new medication, finding they experienced side-effects, and being prescribed another three-month supply of an alternative medication.

The introduction of more significant patient charges had a further impact on the relationships between pharmacists and customers. A visit to a pharmacy began to cost a significant amount of money and pharmacists believed that this changed their relationship with their patients. Pharmacists also had to decide what to do if patients said they were unable to afford their medicines.

The 1991 increase in patient charges — to $15 and $5 — and the lowering of the exemption level brought in a different set of incentives to the pharmaceutical distribution network. Although the consequences of these are not yet clear, some evidence is emerging. Many of the likely consequences of this change are due to the fact that a significant number of prescriptions cost less than $15. In the year to the end of March 1989 the average cost of prescriptions was $18.45 (Department of Statistics, 1990). Few prescriptions cost less than the $5 charge that used to be in force. When the patient contribution was $5 an individual had to have 25 prescription items in a year in order to qualify for free medicine. Therefore most had to pay $125 to reach the exemption limit.

A lower exemption limit (10) and a higher co-payment ($15) changed this. A significant proportion of prescription items cost less than $15, so it has become possible to reach the exemption limit without paying $150. If, for example, a patient is prescribed 10 courses of a cheap drug, say about $5, then they would only pay $50 before reaching the exemption limit. This means that patients have an interest in getting 10 prescriptions for low-cost drugs, and then reaching the limit. Since it is difficult to arrange to get 10 sicknesses which are cheap to treat early in the calendar year, patients may put pressure on doctors, and doctors may suggest to patients whom they see as needy, that they mimic this pattern by writing larger numbers of individual prescriptions for small quantities of drugs — so that patients reach the 10-item prescription limit early without paying much. In a sense, the private judgements of doctors are acting as a form of informal means-testing. This would have a different effect on pharmacies than the stat prescribing encouraged by the $5 and $2 charges. More prescriptions, no matter how large they are, mean more dispensing fees for pharmacists.

This example clearly illustrates the dependence of pharmacists on the prescribing patterns of general practitioners, and the unexpected consequences of the complex system of payment for pharmaceuticals. Clearly, policies introduced to contain the cost of pharmaceuticals can have the opposite effect, depending on how the various actors in the system respond to the new set of incentives.

Now that a significant number of prescriptions cost less than the patient contribution ($15), they do not receive any Government subsidy. This means they do not go through the normal prescription pricing process. It also means that they do not form part of the information base needed by the Health Department for containing the cost of pharmaceuticals. This information would allow the Department to monitor the causes of increased expenditure, and to evaluate the results of attempts to reduce expenditure. Hence another, unexpected result of the changes has been to destroy the comprehensive information base that at one time resulted from the distribution and payment system.

The fact that many prescriptions are now unsubsidized means that pharmacies may vary in the prices they charge for dispensing them. This allows price competition between pharmacists. In the US people often get quotes from various pharmacies before they get prescriptions dispensed. 'Shopping around' may become more common in New Zealand also. Whether this is a good development depends in part on one's view of the role of pharmacists. If they are seen as purely distributors of medicines, then it makes sense to make them compete in order to lower prices. If they are seen as primary health care providers, concerns about continuity of care may arise, as it may be considered desirable for people to go to the same pharmacy each time so that they can build up a record of the person's medication, and identify possible drug interactions, mistakes in dosage, and so on. If this view is subscribed to it may be necessay to concentrate on finding other ways to contain costs.

Case Study II: Generic Substitution

In 1988 the Minister of Health, David Caygill, announced that the Government intended to allow pharmacists to substitute generic drugs for branded drugs. This required a change in the Medicines Regulations and in the pharmacists' code of ethics (Department of Health, 1988). This was not carried out before the change of Government in 1990. However, the issue of generic substitution does raise some interesting questions about the relationships, and the division of labour, between pharmacists and doctors.

As explained above, at present New Zealand pharmacists cannot substitute unless they have an agreement with the prescribing doctor. However, some doctors already give pharmacists written permission to substitute whenever they feel it is appropriate (Moffett, 1987). If the proposal for generic substitution had been carried out, pharmacists would always have been able to substitute except when doctors specifically *prohibited* it.

As well as eliciting a storm of protest from the manufacturers of brand-name drugs who do not want to lose their market share (see the chapter by Lexchin and Norris), generic substitution alters the division of labour between pharmacists and doctors. Pharmacists would have more control over what brand of drug is dispensed.

One would expect that pharmacists would like such a move as it would increase their ability to use their professional discretion. The notion that pharmacists should have more input into the choice of drug therapy is an important part of pharmacists' arguments that they are 'drug experts'. In fact, some pharmacists do see it this way. As one pharmacist said in a personal interview:

Generic substitution is a role that I think pharmacists should be involved in because professionally there's no reason why they shouldn't be. If I can't detect [problems] and be responsible for the drugs I dispense, then I ought not to be classed as a professional.

Doctors, on the other hand, might be expected to oppose generic substitution, in part because they might feel they lose some control over therapy. The pharmacist would become the person who keeps a record of exactly what drug the patient has taken.

In the US anti-substitution legislation had the support of pharmacists when it was introduced in the 1950s to prevent counterfeiting, but this problem subsided in the 1960s. In 1970 the American Pharmaceutical Association committed itself to seek repeal of the anti-substitution laws 'to allow pharmacists more autonomy in drug product selection'. Medical organizations and pharmaceutical manufacturers opposed this move, citing an 'invasion' of the patient-physician relationship and interference in the physician's ability to practice medicine (Wertheimer, 1980). In fact, evidence from the US suggests that once substitution is allowed, the degree to which doctors object depends on what they have to do to indicate this. This is described in the chapter by Lexchin and Norris.

For various reasons there seems to be little consistent support among New Zealand pharmacists for an increased ability to substitute generically. In a recent survey of pharmacists in Otago and Southland, pharmacists indicated that they would rather choose the brand to be dispensed less often than they do at present. Along with informing patients about bureaucratic changes in the health sector, this was the only task which pharmacists wanted to do less of. In the latter case they wanted someone other than themselves or doctors to have to do this, but they wanted doctors to be more responsible for choosing the brand to be dispensed (Roseman, 1989).

There are a variety of reasons why pharmacists might not substitute as often as they could. Pharmacists vary in their opinions about the safety and bioequivalence of generics and in their views of generic substitution. Some also have concerns about liability.

Economic incentives may also affect pharmacists' willingness to substitute. These depend on whether pharmacists are paid by a mark-up or a fixed fee per prescription. Mark-ups can be applied to the particular brand the pharmacist dispenses, or to the cheapest brand. In the former case, pharmacists get paid more for dispensing a more expensive brand. This incentive is avoided if the mark-up only applies to the cheapest brand; they then get paid the same amount regardless of which brand they dispense. This also happens when pharmacists receive a fixed fee per prescription. In Iowa, researchers found that generic substitution by pharmacists increased when their method of earnings was changed from a mark-up (that is, a percentage of the drug cost) to a fixed fee per prescription (United Nations Conference on Trade and Development, 1981).

In New Zealand the situation is complicated by rebates and discounts which pharmacists get from wholesalers. These affect the price pharmacists pay for various brands of drugs, and therefore their incentives to substitute. Co-operative wholesalers are owned by groups of retail pharmacists who distribute their profits as rebates and discounts to members. Discounting is widespread among both these and other wholesalers (Newell et al., 1980). When pharmacists receive a discount on a certain brand of drug, in effect, they receive a higher mark-up when they dispense that drug.

Another factor which affects whether pharmacists substitute is again the issue of stock. Substitution allows pharmacists to stock only one or two brands of each medicine (only one when the medicine is under patent protection) (Moffet, 1987). However, substitution sometimes means extra work in explaining the change to customers.

Again, this case study shows how the commercial and professional relationships in which pharmacists are involved can affect the success of policies intended to contain the costs of pharmaceuticals.

Conclusion

As this chapter has shown, formulating and implementing policy in the area of pharmaceuticals is very complicated. The set of relationships between professional groups, the industry, and the State are complex and laden with the burden of history. Policies can have unintended consequences, and whether these are desirable depends on one's broad view of the appropriate role of each professional group, and on one's definition of the public interest. There are two competing views of the role of pharmacists. The traditional one, in which pharmacists are seen as simply distributors of medicines suggests that immediate attempts to cut expenditure in this area may be appropriate. Many pharmacists, however, are promoting a different view of their role. They believe they have an important role to play in improving and reducing the use of medicines. If we accept this and want to encourage its development then policies should be scrutinized with these goals in mind.

I am completing a PhD on the changing role of pharmacists in the Department of Sociology at Victoria University of Wellington. I am particularly interested in the impact of technology and government policy on the occupation of pharmacy, and the development of clinical pharmacy.

References

Ashton, J. (1991) Health care in New Zealand — looking for a lasting solution *Health* 40: 4-10.

Broadhead, R. and Facchinetti, N. (1985) Drug Iatrogenesis and Clinical Pharmacy: The mutual fate of a social problem and the professional movement *Social Problems* 32: 425-37.

Brompton, S. (1988) How a pharmacist helps cut drug costs *New Zealand Pharmacy* 8: 32-3.

Cammell, J. (1990) Greater Participation at ward level now *New Zealand Pharmacy* 10: 25-7.

Campbell, R.K et al. (1989) Compliance with Washington State's Professional Practice Regulations: 1974 vs 1987 *American Pharmacy* NS29, 5: 42-8.

Committee of Inquiry of the Nuffield Foundation (1986) The Report of a Committee of Inquiry appointed by the Nuffield Foundation *Pharmacy*.

Combes, R. (1981) *Pharmacy in New Zealand: Aspects and Reminiscences* Auckland, Ray Richards.

Deno, R.A., Rowe, T.D., and Brodie, D.C. (1966) *The Profession of Pharmacy: An introductory textbook* 2nd edition. Philadelphia, J.B. Lippincot Co.

Department of Health (17 November 1988) Memorandum for the Minister of Health: *Medicines Legislation* 140/1/3/3.

Department of Health (1991) *1991 Changes to Prescription Charges* Leaflet available in general practitioners' offices.

Department of Statistics (1990) *New Zealand Official 1990 Yearbook* 94th edition, Wellington, Government Printing Office.

Department of Trade and Industry, Department of Health, and The Treasury (6 December 1985) 'Officials' Report on Restrictions on the Ownership of Pharmacies.

Eaton, G. and Webb, B. (1979) Boundary Encroachment: Pharmacists in the clinical setting *Sociology of Health and Illness* 1: 69-89.

Harrison, G. (1990) *Computerised Health Information Systems: A Case Study Analysis of Planning and Implementation* Diploma in Community Health Project, Wellington, Wellington School of Medicine.

Holloway, S., Jewson, N., and Mason, D. (1986) Reprofessionalisation or Occupational Imperialism?: Some reflections on Pharmacy in Britain *Social Science and Medicine* 23: 323-32.

Moffett, W. (1987) Substitution: Name dropping can offend *New Zealand Pharmacy* 7: 6-15.

Newell, K. et al. (1980) The Economics of the distribution of Prescription Medicines in New Zealand *Community Health Studies* 4: 94-103.

New Zealand Hospital Pharmacists Association (January 1990) Submission on the Pharmacy Bill 1989.

Pharmacy Guild of New Zealand (1989) 57th Annual Report and Statement of Accounts.

Pharmaceutical Society of New Zealand (4 November 1988) Submission to the Review of the Pharmacist Occupation.

Roberts, M. (1987) The changing nature of pharmacy in the New Zealand Health Care System *New Zealand Medical Journal* 100: 522-4.

Roseman, P.D. (1989) *The role of the Community Pharmacist and the General Practitioners as seen by themselves and each other, and the interactions between them* B.Pharm.Hons thesis, Dunedin, University of Otago.

Scott, C., Fougere, G., and Marwick, J. (1986) *Choices for Health Care: Report of the Health Benefits Review* Wellington, Health Benefits Review.

Scott, G. and Gini, P. (April 1991) A note on the demand for prescriptions: presented at the Wellington Health Economists meeting.

Tilyard, M., Dovey, S.M., and Rosenstreich, D. (1990) General Practitioners' views on generic medication and substitution *New Zealand Medical Journal* 103: 318-20.

United Nations Conference on Trade and Development (UNCTAD) (15 December 1981) *Trade marks and generic names of pharmaceuticals and consumer protection* Report by the UNCTAD secretariat p. 17-18 TD/B/C.6/AC.5/4.

Wertheimer, A. (1980) Editorial: The Irony of Drug Product Selection *American Journal of Public Health* 70: 473.

Working Group on Occupational Regulation (1970) Report on The Pharmacy Act.

Pharmaceutical Costs and Regulation: From the Minister's Desk

Helen Clark, MP

I served as Minister of Health from February 1989 until October 1990. Throughout that time pharmaceutical-related issues were numerous and complicated. Conflicting and vested interests kept those issues in the headlines. The public interest was never their concern. Profit and clinical freedom at public expense often were.

Counterweights to those powerful interests were few. Latterly the Public Health Association (PHA) emerged as a prominent voice, attracting, of course, derision and scorn from the pharmaceutical industry. Public-spirited individuals such as Professor Laurence Malcolm of the Wellington School of Medicine spoke out. The Health Department's voice was more timid than might be expected of a regulator established by statute to act in the public interest.

Government Efforts to Contain Pharmaceutical Costs

Growth in State pharmaceutical expenditure in the 1980s was of considerable concern to the Government. Drugs had been freed from price control by David Caygill in his capacity as Minister of Trade and Industry. Both he and Michael Bassett before him as Minister of Health had encouraged a more hard-nosed attitude by the Health Department towards price negotiation with pharmaceutical companies. The Government was under considerable fiscal pressure, given that the 1988 cuts in taxation were producing less revenue than spending commitments demanded. Pharmaceutical bills in excess of half a billion dollars to the State were an obvious target for expenditure reduction. Both the Minister and the Department of Health redoubled their efforts to that end.

The public interest in that activity was threefold. Firstly, the taxpayer should not be paying unnecessarily high bills for pharmaceuticals. Dollars wasted in this manner could have been better invested in health or elsewhere. Secondly, some of the waste undoubtedly arose through prescribing which was either unnecessary *in toto* or excessive in amount. Thirdly, the public had an interest in seeing the Department's agents pushing a hard bargain on the price of a class of drugs, so that there could be a suitable product on the Drug Tariff free of charge to the consumer. Part-charges on top of prescription charges were always resented and resulted in a not inconsiderable volume of ministerial correspondence which almost invariably suggested a lack of Government commitment to the public health system.

The pharmaceutical industry, not surprisingly, argued differently. Government tactics, they alleged, were selling the consumer short. Drugs available to consumers elsewhere were either being withheld from New Zealanders altogether or they had a punitive part-charge placed on them. Driving down company profit margins would lead to a diminution of funds for research into new 'life-saving' drugs. These arguments assumed a dreary, self-serving familiarity over time.

This chapter will sketch the outline of a selection of the pharmaceutical issues which arose during my 21 months as Minister of Health. In most cases heavy and critical lobbying from the pharmaceutical industry was encountered. Sadly, in some, medical practitioners who should have known better joined their cause without taking into account the broader picture. In the process, a good deal of misinformation was disseminated.

It should be made clear at the outset that while Government's fiscal requirements gave impetus to planning ways of reducing the State's outgoings on pharmaceuticals, there were also sound health reasons for examining the level and nature of prescribing which had direct consequences for public expenditure. The level of prescribing for hypertension was certainly open to question; as was that of benzodiazepines, particularly for the elderly with medium- and long-term negative effects. High cholesterol levels also appeared likely to become a condition treated invariably by pharmacological means unless medical practitioners could be encouraged to address the causes and not the symptoms of the problem. Yet the inducements to prescribe, and to prescribe new and expensive drugs, were many, as has been ably outlined in the report of the PHA workshop on pharmaceutical advertising and promotion (Kawachi, 1990).

The lavishness of the drug industry's advertising in publications aimed at medical practitioners indicates the level of financial return likely to flow from a drug achieving popularity among prescribers. Profit is, not unnaturally, the primary objective of the industry.

Strategy for the Reduction of Drug Expenditure

A multi-faceted pharmaceuticals strategy was required to deal with the many objectives of the campaign to reduce drug expenditure and unnecessary usage. The Department of Health drew up the outline of that strategy for me in July 1989. Sixteen different projects were outlined, to be worked on over the short and medium term (see Table). It says something for the complexity of the task, and perhaps of the Department's capacity to organize in pursuit of it, that too few fruits of the exercise, even with respect to those parts of it which were well publicized, have been seen to date. For a time I required weekly reports on progress, but as my attention was often diverted to other issues, over time the reports ceased being written and the impetus to achieve the goals diminished.

Table: Projects in Health Department Pharmaceuticals Review Strategy: July 1989[1]

A. *Projects aimed at encouraging greater competition*
   Generic Substitution.[2]
   Parallel Importing.
   Pharmacy Bill.

B. *Projects aimed at influencing prescriber behaviour*
   Treatment Protocols.
   Preferred Medicines Lists.
   Prescriber Education.
   Marketing Activities of Pharmaceutical Companies.

C. *Projects aimed at opening up choice to the consumer*
   Medicines Classification: Increasing the Range of Over-the-Counter Products and the Reclassification of 'Pharmacy Only' Medicines.
   Patient Education.
   Generic Substitution.

D. *Projects examining the Drug Tariff*
   Review of the Structure of the Drug Tariff and the Process of Adding Medicines to the Tariff.
   Impact of New Products on the Pharmaceutical Benefit Scheme.

E. *Other Miscellaneous Projects*
   The Impact of Demographic Changes.
   Volume Prescribing.
   Auditing of Returns from Pharmacies for Reimbursement under the Pharmaceutical Benefits Scheme.
   Doctor Availability.
   Changes to the Customs Tariff and Import Duties.

1. The groupings of projects have been devised by the author.
2. Generic Substitution appears in two groupings.

The majority of the projects fell into the categories of encouraging greater competition, influencing prescriber behaviour, opening up greater choice to the consumer, and the admission of products to the Drug Tariff and the implications of that.

Encouraging Greater Competition

The projects in this category focused on drawing up regulations and policy for generic substitution, law changes and mechanisms to enable the parallel importing of pharmaceuticals by the Crown, and policy on and drafting of the contents of the Pharmacy Bill. They were to absorb many years of Departmental officers' time. They were potentially highly controversial because of their impact on vested interests.

None of these projects has yet been followed to completion. The Pharmacy Bill lies before a select committee, stalemated until the National Government makes policy decisions on the extent of deregulation of retail pharmacy it is prepared to allow. Not a single aspirin, let alone any more sophisticated drug, had been imported by the time I left office, nor by the time of writing, despite amendment of the Medicines Act to facilitate this. To date generic substitution has not proceeded either. I return to generic substitution and parallel importing later in this chapter.

Influencing Prescriber Behaviour

Projects in this category sought to encourage cost-effective prescribing and have generated some useful work. As prescribers in New Zealand have obtained their postgraduate information about pharmaceuticals largely from the drug industry, it would not be surprising if consideration of cost to the taxpayer, and objective measures of efficacy, had little influence on their behaviour. Clearly there was scope for ongoing prescriber information from disinterested bodies. One project proposed to look at how that information might be best conveyed to prescribers, and another at the concept of a voluntary or mandatory regional or national preferred medicines list.

Experience with the Nelson Prescribing Study from 1988–1990 in which almost all general practitioners in that region worked together to draw up, implement, and evaluate a preferred list, suggested that a collaborative effort with the profession would be likely to produce the best results. For that reason also I later determined in 1990 that the Department should fund what became known as the Preferred Medicines Co-ordinating Centre at the Wellington School of Medicine rather than attempt to undertake the task of formulating and promoting preferred lists itself. Both those projects are ongoing, although one questions whether the State will continue to have any direct incentive to finance and encourage them as it reduces its funding for the Pharmaceutical Benefits Scheme and passes the costs over, increasingly, to users themselves and/or to their insurers.

Another project in this category was to encourage the development of treatment protocols for common conditions. New Zealand has had few clear national guidelines for such conditions, with the result that pharmacological treatment is widespread but of doubtful cost benefit or benefit to the consumer. A taskforce has been working on a treatment protocol for hypertension for some time. Raised cholesterol levels and conditions which presently lead to considerable prescribing of benzodiazepines would also seem obvious targets for such attention. Development of such protocols requires the participation and input of professionals. In the end they must be prepared to stand behind them if the onslaught of criticism from the industry, which stands to lose profits from lower prescribing, is to be thwarted.

Finally, in this category, the marketing activities of pharmaceutical companies, so often aimed directly at the medical profession, were to be reviewed to assess whether they should be more tightly controlled. The PHA made a very useful contribution to this project, convening a workshop on the issue in March 1990 funded by the Department of Health, publishing its proceedings, and producing in the outcome a very useful source document for policy development (Kawachi, 1990). Through this and other activities, however, it attracted the opprobrium of the so-called Researched Medicines Industry (RMI), the industry association of pharmaceutical companies in New Zealand. The RMI approached the National Opposition about ceasing the Governmental grant from Vote: Health to the PHA should the Government change. The Opposition's response must have been music to the industry's ears. At that time the MP for East Coast Bays, Mr McCully, was assisting Mr McKinnon, Opposition health spokesman. Mr McCully's opinion, as reported by the RMI's Chief Executive in leaked minutes of a meeting, was that 'the funds made available to the PHA were a waste of money and needed to be looked at' (RMI, 1990). In National's first Budget in 1991, funding to the PHA was duly removed.

Opening up Choice to the Consumer

The third category of projects involved consideration of how to empower consumers with knowledge about the appropriate use and non-use of medication and give them some control over the cost of medication when used. For many, a visit to the doctor is considered unsuccessful if a prescription is not written. The project on public education about drug and non-drug therapies fitted well with other health initiatives, such as the Health Charter and its associated New Zealand Goals and Targets, released in December 1989 (Clark, 1989). This contained a heavy emphasis on health promotion and the strong message that the key to better health for most of us lies in our own hands. As the Minister was often wont to say, good health is not derived from the doctor's surgery, the chemist's shelves, or the hospital bed!

Another project in this category proposed the reclassification of some medicines so that they could be more readily available to the public. There seemed little reason why a number of medications remained available only on prescription. That meant extra cost for the public who had to pay doctors' fees, and for the taxpayer who subsidized them for consultations which in a number of cases the pharmacist was well qualified to undertake without charge. A proposed list of reclassifications was drawn up in mid-1989 and presented to representatives of the medical and pharmacy professions, consumers, the RMI, and others for comment. Both the medical and pharmacy representatives present made strenuous objections, which were repeated in later submissions, to aspects of the proposals. As the Department noted in written advice to me: 'It is difficult to separate out the concern of the doctors and pharmacists for the public health from their concern for their income'! (Department of Health, 14 July 1989).

In order to work through the recommendations logically I determined that the Medicines Classification Committee should be reconvened for that purpose. Now, at the time of writing in late 1991, the outcome of that work is to hand, but the public will continue to pay doctors' fees unnecessarily for access to those medicines which should be more generally available.

Generic substitution plans also sought to empower the consumer with respect to medicine costs. The Department's practice was to set the price for a class of drugs at the minimum level for which any one product in that group could be obtained. Other drugs of the same type where the company insisted on a higher price then carried a part-charge at the point of dispensing to the consumer. Generic substitution would have enabled the pharmacist to offer the consumer the option of having dispensed a drug bioequivalent in its composition to that prescribed by the medical practitioner. Where the latter carried a part-charge to the consumer it was clearly in his/her interest to have that choice. Nothing, however, was simple about generic substitution, as I shall explain later in this chapter!

The Structure and Content of the Drug Tariff

The fourth major category of projects envisaged in the 1989 strategy dealt with the structure of the Drug Tariff, the process of adding medicines to it, and the impact of new products thus added on the cost of the Pharmaceutical Benefits Scheme.

Recommendations on the admission of a new medicine to the Drug Tariff are made by the Pharmacology and Therapeutics Advisory Committee (PTAC). Prior to that, however, the Minister must make a decision on whether to approve a medicine for distribution in New Zealand. That decision is taken after receiving advice from the Medicines Assessment Advisory Committee (MAAC). While a medicine is under review through these processes, it is not unknown for manufacturers, often in collaboration with clinicians, to work to create a demand for the product. The public is left as a pawn in a larger game. Once the medicine is admitted to the Drug Tariff, sales to the public, and cash-flow to the company, are virtually assured.

By 1989 the Department of Health was ready to recommend to the Minister that new products should not be as easily admitted to the Drug Tariff as in the past. They advised that studies in the United States and the United Kingdom 'have indicated that only a small percentage of new products are fully innovative. In the United Kingdom study, 65 per cent of the 103 new products were judged non-innovative' (Department of Health, 14 July 1989).

New Zealand appeared to have acquired the reputation of being a relatively easy country in which to register new products. That raised the question not only of the extra expense which might be generated by paying for newly-marketed versions of old remedies, but also the question of whether the public might be exposed to unnecessary risk by too ready acceptance of drugs not yet readily available elsewhere. Logic would suggest that New Zealand had fewer resources than most other Western countries to evaluate the efficacy and safety of drugs. Our Accident Compensation Scheme of no fault reimbursement for injury due to misadventure also made New Zealand an attractive market for new products.

The initial step taken to deal with this set of problems prior to my appointment came when the Department notified the MAAC that new medicines would not be referred to it for advice unless they had already been approved for sale in another Western country which had an acceptable regulatory system. Members of the Committee affected professional indignity in response and threatened to resign. The Department asked that I advise the Committee that as a general rule I was unlikely to be satisfied with the safety and efficacy of a new product without evidence of the results of large scale use following approval in a major Western country. That I was happy to do. The Department also noted that some applications might be sent to the Committee in advance of such evidence. In practice they were, but I recall on occasion having positive recommendations for such products put before me which I declined to approve because of my concern that New Zealand with its limited capacity to judge safety and efficacy might lead the way in approval of a new drug, and thereby give the company concerned a lever with which to pressure other countries' regulatory authorities.

In retrospect I regret that I did not ask members of both committees concerned with the evaluation of medicines to declare any interests they may have had with pharmaceutical companies. That request was made of members of the Medicines Adverse Reactions Committee (MARC) and is, I believe, an important discipline for members to submit themselves to. The smallness of New Zealand makes it near impossible to appoint expert committees with members free of present or past associations with pharmaceutical companies.

The major pharmaceutical issues demanding my attention in 1989 and 1990 were generic substitution, parallel importing, high-cost medicines, and the safety of the asthma drug, fenoterol.

Generic Substitution

The New Zealand Government was not alone in proposing to enable generic substitution by pharmacists. Nor was New Zealand alone in being bombarded with mischievous propaganda by the drug industry which seeks above all to maintain market dominance for its brand-name products. The irony of generics often being produced by the very same companies whose propaganda blackens them will escape no one. Clearly, profit margins are at stake, not quality and safety.

My predecessor, David Caygill, initiated moves to draw up regulations under the Medicines Act to enable generic substitution by pharmacists where the client agreed. From the outset the pharmaceutical industry in New Zealand sought to discredit the move. In October 1988, the Director General of Health, Dr George Salmond, released a press statement in response to those attacks. He said that:

Recently reported comments about generic products being 'potentially lethal' are mischievous and serve only to undermine public confidence in the pharmaceutical industry and regulatory controls . . . I believe it is not without coincidence that these claims are being made at a time when the government is expressing serious concern about the escalating drugs bill and looking at ways of improving the return on the taxpayer's dollar invested in this area. The public should not be muscled into thinking a cheaper version of a medicine is necessarily of lesser quality; it simply reflects the ability of some companies to either manufacture more efficiently or take more reasonable profits (Salmond, 1988).

The industry kept up the pressure both in public and in private. On a Sunday morning in March 1989 it used television time to broadcast directly to the medical profession and the public. In a letter notifying me of the broadcast, the RMI's Executive Director advised that it would examine 'proposed amendments to the Medicines Regulations 1984 and associated ethical issues including the effect of generic prescribing, adverse effects following substitution, professional response to changes in legislation, and other matters' (Hardy, 1989).

Representations from the RMI also went directly to the Minister of Overseas Trade, the Prime Minister, and the Opposition in attempts to undermine the Minister and the Department of Health's stance in support of generic prescribing.

The Opposition appeared to enjoy direct links to the pharmaceutical industry through a personal connection in its leader's office. It was a willing conduit for industry inquiries on many occasions through the medium of parliamentary questions. Those questions were so detailed and precise that they could have been written only by a person very close to the issues and with some inside knowledge of issues arising within the Health Department as plans for generic substitution were being developed.

Indeed, a key part of the industry's campaign against generic substitution was to seek to discredit the Departmental advice to the Minister. These attempts were brazen. My one and only meeting with the RMI took place on 13 June 1989. In a letter sent to me from the RMI's new chief executive, Mr W.P. McLauchlan, on 7 June 1989, he advised that the first issue the delegation wished to raise was generics registration and substitution, and stated: 'The R.M.I. is concerned that the Minister is receiving inaccurate information from the Department of Health . . .' (McLauchlan, 1989).

Similar views were expressed in their submission to the Minister of 24 January and again in the letter of the Chairman, Mr Main, to the Minister on 13 March.

My recollection of the meeting is of its unpleasant nature as RMI representatives sought to cast slurs on the integrity of the Departmental officers who worked on pharmaceutical issues. I had no reason whatsoever to doubt their integrity. Under persistent attack from the industry, however, it seemed, understandably, that some developed a siege mentality which resulted in an inability to acknowledge difficulties which might be present or the inevitable mistakes which might be made from time to time.

The industry's letter to the Prime Minister of 15 February 1989 assumed a particularly threatening tone. At that time the industry association was still named the Pharmaceutical Manufacturers Association (PMA). The Researched Medicines Industry title emerged shortly thereafter in an attempt, one assumes, to give the industry association greater credibility and status and the appearance of being professional and expert.

The President of the PMA, Mr G.T. Bethell, invited the Prime Minister to consider what might be:

(a) The view of a senior multi-national corporate executive considering his corporation's involvement and investment in New Zealand and what might be the action resulting from this.
(b) The impact on the New Zealand elector. . . . Taking into account all the factors listed above it could well be anticipated that a decision could be taken to cease further investment in New Zealand and not to support the registration of new products here (Bethell, 1989).

The letter went beyond issues of generic substitution to complaints about the Department of Health's policy not to consider new medicine applications until the product had been registered in other major Western countries. It claimed that if it were not possible to gain early registration in New Zealand, few companies would conduct research here and that 'consequently clinicians would be deprived of the intellectual stimulus and satisfaction involved with research'. The tale of woeful consequences of Departmental policy then flowed on to include the likelihood of less industry investment in medical education, the absence of newer and more effective medicines, and a shrinkage of the range of dosage forms available.

An equally dire picture of the effects of generic substitution was also painted:

Legalized substitution will result in large numbers of patients experiencing serious side effects and complications due to differing efficacy between brands. Patients will lose confidence in their doctor, in medicines and in the government. Ultimately the range of medicines available in New Zealand will be more akin to those found in a developing country. This will cause increasing public concern, frustration and loss of confidence (Bethell, 1989).

In April the RMI demanded publicly that the Minister of Health should immediately withdraw eight generic drugs from the market. It claimed that people's health was at risk from the products said to be taken by 25,000 New Zealanders each day. It alleged that the same products had been withdrawn from the Australian market the year before (RMI, 1989).

Unfortunately for the RMI its allegations were well wide of the mark. As the Department of Health pointed out in its 21 April rejoinder:

The medicines recalled in Australia are made in that country and to an Australian formula. The medicines named by the Association have been sold in New Zealand for periods of up to ten years without causing problems. They are made in a New Zealand factory to a New Zealand formula (Department of Health, 21 April 1989).

My enthusiasm for proceeding with regulations to permit substitution was unabated by the industry campaign. The latter had a sense of *déjà vu* about it to anyone familiar with industry tactics abroad. It did seem important, however, to have the support of the Medical Association for Government proposals. The Department of Health was asked to work with it on any outstanding issues where the Association may have had doubts. Further consultation was productive in overcoming what could otherwise have been damaging medical opposition to substitution.

Departmental work on the regulations required for implementation of substitution proceeded, although always, it seemed, at a snail's pace. By November 1989 the Department had turned its mind to the need for information for doctors, pharmacists, and the public. Pharmacists were agreeable to making available a Department of Health leaflet on substitution in their pharmacies. Work was done on the wording. Then, in anticipation of the likely industry assault, television advertisements were made featuring Billy T. James talking about the purpose of substitution and its benefits. I was not prepared to put the regulations up to Cabinet and the Executive Council for approval until I was confident that all possible steps had been taken to ensure smooth implementation.

What I did not know was that behind the scenes in the Department there appeared to be some divergence of opinion on whether bioequivalence, that is, equivalent physiological effect for equivalent preparations, could be guaranteed for all generics available in New Zealand. Opposition parliamentary questions at the instigation of the RMI had sought assurances about the Department's requirements and procedures for assessing bioequivalence. Earlier ministerial experience had made me sceptical about accepting at face value Departmental advice on technical matters on which I had no expertise. My answers to such questions therefore were generally prefaced with the caveat 'I am advised that . . .'.

That turned out to be a sensible precaution in the case of bioequivalence. I was alerted to the possibility that the path forward to substitution was not as clear as I had been led to believe, only after an Official Information request brought relevant Departmental files to the fore. File notes indicated internal Departmental disagreements on whether claims of bioequivalence for certain drugs could be supported. On 1 June 1990 I wrote to the Director-General outlining my concern that generics tested in earlier years for bioequivalence may not have been tested to a standard which would allow me to state publicly that those being substituted were actually equivalent. I said that I was not prepared to proceed with substitution unless I had full confidence in the Department's testing procedures.

In response the Department stated that prior to 1989 the evaluation process for both generics and original brands was inadequately documented. I had received that advice as early as March 1989, but had never been advised that that could raise doubts about bioequivalence. Now the Department added that caveat, noting that in the absence of adequate documentation it would be difficult to defend the decisions previously reached.

The options then became to proceed regardless, abandon substitution, or take a step-by-step approach whereby generics could be placed on a substitution list as each medicine was reviewed. The latter was the only option. By the time I left office, the regulations were still unpromulgated, not because of the success of industry campaigning, but because the Department in the final analysis had not provided the necessary assurances to make them publicly defensible. No doubt this left the RMI in a position to relitigate the substance of the issue with a new government more friendly towards it. There is still no sign of the regulations.

Parallel Importing

The saga of legislation to permit the parallel importation of pharmaceuticals by the Crown has similar ingredients of industry manœuvring and less than transparent bureaucratic advice. Parallel importing was one of a number of strategies to be pursued in the campaign to reduce pharmaceutical spending by the State. Some 80 per cent of prescription medicines used in New Zealand were imported. In the Department of Health's view, a number of them were sold at prices considerably above those applying elsewhere.

The pharmaceutical market has special characteristics, such as the need for public safety and protection, which require controls to be placed on its operation. Unleashing market forces could undoubtedly lead to cost savings, but also has the potential to compromise the quality of the product available to the public. The option of allowing selected parties to trade in importing pharmaceuticals in competition with existing suppliers was considered, but judged unlikely to result in savings to the State. The option of allowing the Crown to intervene as the sole parallel importer had the most potential for fiscal savings and could also guarantee the safety and quality of the product.

The law has required that importers produce a manufacturer's test analysis certificate to satisfy safety standards. In practice, manufacturers monopolize supply to the New Zealand market through their own agents by preventing other parties having access to the information required to satisfy New Zealand's safety standards.

The Cabinet agreed that legislation to permit the Crown to import pharmaceuticals should be introduced and passed on Budget night in July 1989. The legislative option chosen was blunt. It simply exempted the Crown from the provisions of the Medicines Act 1981 when importing and selling medicines. The Crown's integrity and practice as a guarantor of pharmaceutical safety standards was to be relied on for product safety. It has the ability to test medicines against specifications and would have done so when it imported. Potential savings of around $50 million annually to the State were identified by the Department.

Concurrent with the preparation of the amendment to the Medicines Act, the Department readied itself for importation. It set up its own trading company, and made arrangements to source products through an Australian hospital organization. It advised me prior to the Budget that it would be ready shortly to fill orders for local wholesalers. It also advised that reaction to the move was unlikely to be muted.

In fact the initial reaction was muted. The public was unlikely to be concerned at moves to lower the price of drugs if the quality remained the same, and so unlikely to be moved by industry special pleading. Industry tactics therefore were different. Most pressure came from abroad through representations from foreign governments, particularly that of the United States. The inevitable planted parliamentary questions from the Opposition followed. The Ministry of External Relations and Trade's submissions to Government via cable traffic from Washington showed considerable agitation.

In a cable from the New Zealand Embassy in Washington on 10 August 1989, the means by which the international pharmaceutical industry was working to undermine the New Zealand initiative were made clear. The RMI in New Zealand had contacted the American Pharmaceutical Manufacturers' Association (APMA), which in turn had filed a formal letter of complaint with the United States Trade Representative. According to the cable the APMA's complaint was that 'the New Zealand legislation would seriously undermine the TRIPS [Trade Related Intellectual Property] negotiations currently under way in the GATT' [General

Agreement on Tariffs and Trade] . . . and asserted that 'apparently New Zealand had now fallen in with countries subject to special 301 provisions of the omnibus Trade Act' (New Zealand Embassy, 10 August 1989). The effect of that assertion appeared to be to suggest that the United States Government should take retaliatory action against New Zealand trade. Such action would of course be unlikely in practice because of the limited number of trade restrictions applying in New Zealand in relation to American products.

In the view of New Zealand diplomatic staff in Washington, the Americans were low-key and non-confrontational about the matter and preferred to resolve the matter 'without having to resort to the instruments which were available to them'. The cable went on to say, however, that the APMA was 'a powerful lobby group' and was likely to 'use every opportunity to appraise Congress of its concern, should the matter continue to be seen as being unsatisfactorily resolved'.

The United States Trade Representative's office also advised the New Zealand Embassy that the APMA had contacted its counterparts in Switzerland and the United Kingdom and that a reaction from those countries could be expected.

In Wellington officials from the Justice Department and Ministry of Commerce also expressed concern. It seemed that the broad brush nature of the amendment had at least the potential to place New Zealand in breach of international conventions and agreements to which it was party. The Customs Department was concerned that the Crown as an importer might be evading its own customs duties and goods and services tax.

In retrospect it seems incredible that Health and Treasury officials did not draw to Cabinet's attention the need for consultation with other departments to avoid the pitfalls of a crude amendment to the Medicines Act. Government quickly agreed after the potential problems were drawn to its attention that amending legislation was required to permit parallel importing to proceed smoothly.

Despite the Budget night amendment enabling parallel importing by the Crown, none occurred for a variety of reasons. Arrangements with Australian suppliers appeared to cool. It seemed that the hospital organization which had previously agreed to supply had been subjected to considerable pressure from some pharmaceutical manufacturers. Despite visits to Australia by senior Health Department officials a satisfactory source of supply never seemed to eventuate. When I left office 15 months after the power to import had been granted, the Crown had not imported as much as an aspirin; nor had it when this book went to print.

Two final points about this episode are worth noting. Firstly, the Department authorized some background research into pharmaceutical companies operating in New Zealand. Its inquiries revealed that one prominent company was reporting a profit in New Zealand which was only a small fraction of the profit its parent company reported in the United Kingdom. It appears that transfer pricing and routing profits back to base via tax havens are well-developed techniques in the industry, a matter no doubt of considerable interest to our taxation authorities.

Secondly, pharmaceutical expenditure for 1989 to 1990 was in line with budget targets which represented a drop in real terms. In my view the mere threat of parallel importation played a part in that result, but could not be expected to continue to do so in the long term.

High-cost Medicines

Debate about high-cost medicines involved the public to a greater extent than did any other aspect of the pharmaceutical cost-control exercise. The tactics used by the industry to promote them were dubious. A common technique was for the manufacturer to collaborate with medical practitioners in making the latest 'life-saving', 'wonder' drug available to their clients as part of a clinical trial. The term 'clinical trial' itself is highly misleading as the medicines concerned were invariably trialled elsewhere before arriving in New Zealand. Often a clinical trial in New Zealand seemed to represent little more than a marketing exercise for the manufacturer of the drug and the opportunity for a medical practitioner to experiment with that new drug at company expense.

Problems for the State begin when the clinical trial is deemed to be complete. The manufacturer is invariably seeking the admission of the drug to the Drug Tariff from where it may be supplied at considerable public expense. A company will be looking to recoup the high costs of research, development, and publicity associated with the launch of a new product into the market.

In high profile cases in New Zealand in 1989 and 1990, manufacturers ceased free supply of the drug at the end of the trial. Then, directly or indirectly, public pressure was applied for free supply of the expensive medication. Similar tactics could be observed in the hyperbole surrounding the promotion of selected drugs, variously for cholesterol reduction, for the treatment of clients with renal failure who experience anaemia, and for the treatment of Parkinson's Disease.

Some years earlier, the Department of Health had adopted the stance that manufacturers had a 'responsibility to continue treatment of patients used in clinical trials until the product had been considered by the Pharmacology and Therapeutics Advisory Committee or at least a year after obtaining consent to market'. Only thereafter would Section 99 approvals for supply as a supplementary pharmaceutical benefit be granted. The move was made, the Department advised, precisely in order 'to counter the activities of some companies who [sic] enrolled large numbers of patients on to a "clinical trial", then abandoned them leaving Government to continue the cost of the treatment'. The Department concluded that 'these activities were little more than a marketing ploy' (Department of Health, 25 June 1990).

In fact, that procedure did not stop the practice. In the case of 'Zocor', a cholesterol-reducing drug, a co-ordinated campaign associated with some medical practitioners ensued to pressure Government into funding widespread free supply. A number of those practitioners had been involved in the clinical trials and had switched significant numbers of their clients on to the new drug. I was advised that they also proposed to switch other existing patients from other treatments to Zocor prior to it being assessed by the PTAC which makes recommendations to the Minister.

The issue went to the PTAC on 30 May 1989. It noted that Zocor was 'effective in lowering serum cholesterol levels in patients not responding to existing products', but expressed concern about 'the potential for high expenditure of public funds if Zocor becomes the treatment of first choice for all patients with an elevated serum cholesterol level'. On balance the Committee considered that 'the medicine should be provided where other treatments, including diet, have failed and where other risk factors are present' (PTAC, 1989).

The Committee then set down detailed guidelines for approval of supply of the drug as a supplementary pharmaceutical benefit, only after assessment of an application made by a specialist physician. After further discussion with the Department, the recommendations were approved by me.

Zocor then became a media football, pursued most vigorously by Paul Holmes of Television New Zealand. The style of his coverage of the issue is well summed up in his presentation of 14 November 1989. The transcript records as follows:

Good evening to you. The drug called Zocor. Zocor. It is one of the best and latest drugs for people with heart conditions and for people with high cholesterol. They say it is marvellous, in fact they call it a wonder drug. It has no side effects apparently, very few at least (Holmes, 1989).

It followed that only heartless and mean-spirited ministers and officials could deny this life-saving drug to the public; yet denying it they were, unless stringent conditions could be met. Of most aggravation to Holmes, correspondents, and the Returned Services Association which burst forth months later at its annual conference, was the restriction on free prescription of Zocor for the over-sixties. Arguments that the drug stood to do that age group little if any good were paid no heed. The decision was denounced as ageism.

On the Holmes show and in other media, anecdotes abounded of individuals who had been denied free supply of Zocor and were convinced that the Minister and Department of Health had passed a death sentence on them. Professionals involved in the applications fanned the fires of the dispute.

Rationality is quickly lost in the heat of these debates. The need for cost containment is not a message the public wants to hear about health. Expectations are that a public health system must provide regardless of cost wherever there is need. Cost-effectiveness is dismissed as code for dodging that responsibility. Yet with Zocor the case seemed clear. Departmental advice was that if Zocor was prescribed at the cholesterol level and in the dosage recommended by the manufacturer, its annual cost to the State could reach $200 million. The guidelines established by PTAC were expected to keep expenditure to under $25 million.

Despite considerable public antagonism, the State held the line on Zocor. Some revision of the guidelines by PTAC and the Department occurred which gave added flexibility in assessing applications, and removed the absolute age criterion. Overall, however, the Minister, the Department, and, in the end, the Government sustained considerable damage from industry manipulation of public opinion aided by members of the medical profession.

Fortunately, the campaigns mounted to promote Eprex and Deprenyl (two other high-cost drugs) failed to attract public and media attention to the same degree as had Zocor. In each case, however, a letter-writing campaign by consumers and their relatives was instigated, apparently by doctors involved in the clinical trial. Attempts were made in the case of Eprex (a specialized drug for kidney failure) to bully the Government into obtaining recommendations from PTAC earlier than the normal procedure required. The Department advised that the cost for a year's treatment of each individual with Eprex could range between $14,000 and $22,000, and that annual costs to the State could reach $10 million. The Department was not aware of any other country where Eprex was funded on the Drug Tariff.

In this case the National Kidney Foundation was lobbied publicly for the drug. A statement was issued by Hardie Marks and Associates in

Auckland, allegedly on the Foundation's behalf, claiming that the Minister was 'denying the 500 New Zealanders with kidney failure a drug which can vastly improve their quality of life and ability to work' (Hardie Marks and Associates, 1990).

Kidney specialists and clients were quoted in the statement. The public relations consultants involved in this exercise appeared to be closely linked to, if not the same as, the Health Consulting Group which worked closely with the RMI. In 1989 and 1990 the Group produced *The Health Report* which viciously attacked Government policy on pharmaceutical issues. Hardie Marks and Associates had also released the RMI's statement, cited earlier, on the withdrawal of the eight generic drugs in April 1989.

The Fenoterol Debate

The final pharmaceutical issue to be highlighted in this chapter as attracting a good deal of publicity and debate about Government handling of pharmaceutical issues is the debate about the safety of fenoterol. The issue is well covered later in this book from the point of view of the researchers whose work showed the lethal effects of fenoterol, but these additional comments from the perspective of the Minister's desk are offered.

The Health Department brought the likelihood of publication of the research critical of fenoterol to my attention early in 1989. From the outset it seemed strangely hesitant about accepting the implications of the research. My impression was that sensitivity to its relationship with the pharmaceutical company was being accorded a greater priority than interest in research which indicated that many of New Zealand's asthma deaths could be due to the use of fenoterol.

Certainly the company, Boehringer Ingelheim, was shown every consideration when the matter was under review by the Department and MARC. The Department advised on 22 February 1989 that the Committee's meeting scheduled for 24 February should be postponed while copies of all the papers relating to fenoterol were made available to the company so that it could make a submission to the Committee when the issue was considered. I agreed to this procedure, given the advice that not to allow it would open the Committee's recommendations to judicial review.

The Committee eventually met on 22 March 1989. It concluded that there was no reason to withdraw fenoterol from the market at that time. The Department then prepared itself for publication of the research in the *Lancet* and for the likely public and media interest in the availability of a drug said to be responsible for numerous deaths. Its attitude towards the research was clearly somewhat sceptical. A draft press release prepared by the Department referred to the study associating the drug with users' deaths, and then quoted the Principal Medical Officer for Medicines and Benefits, Dr Risely, as saying: 'This is unproven. If there is a risk with the asthma medicine fenoterol, it is an extremely low risk, (Department of Health, 11 April 1989).

That was an extraordinary statement to make at that time, and certainly appeared to prejudge the issue for the Department.

The Department asked Boehringer Ingelheim to keep confidential the research released to it in order that it could respond to allegations about the drug in its submission to MARC. My office, however, was advised on 4 April by Professor Ralph Edwards, a member of the Committee, that he had been contacted by the West German Adverse Reaction Committee which knew of the critical research. There seems little doubt that there had been a breach of what the Department had believed was an undertaking of confidentiality. Thereafter the researchers became very reluctant to hand information over in confidence to the Health Department because of their belief that it was a conduit to the manufacturer, giving the latter sufficient time to organize counter-publicity and disclaimers about findings detrimental to its product.

MARC met again on 26 April to consider the recommendations of Professors Elwood and Skegg of Otago University on the fenoterol research. The recommendations were that action be taken on the basis that there may be an increased risk of death from asthma among certain patients who were prescribed fenoterol and that current practice in the use of fenoterol for such patients should be modified. They also recommended further research.

The recommendations were endorsed by MARC, which again noted that no immediate action to withdraw the drug was required. The Health Department sent a letter to all doctors and pharmacists based on MARC's recommendations. It noted in advice to me that 'Boehringer Ingelheim have a vigorous media campaign prepared. It is not known when it will be launched' (Department of Health, 27 April 1989).

In fact Boehringer Ingelheim went public before the Department's letters were due for release, thus ensuring that it had the first and favourable word on the subject.

At around this time I departed for Canada and the annual meeting of the World Health Organization, leaving the matter to be handled by the Acting Minister, Dr Michael Cullen. A controversy raged publicly over whether or not fenoterol should be withdrawn from the market and whether the Department had acted appropriately on a number of scores, including the early release of the research findings to Boehringer Ingelheim. It was also of some public interest that fenoterol did not have Food and Drug Administration approval in the United States. A public inquiry into the drug was demanded by the Medical Association. Doctors complained that they still did not have enough information on which to base decisions on its use. For reasons not clear to me at the time, but in retrospect because of sensitivity to Boehringer Ingelheim's opinion, the Department had decided not to send the abstract of the researchers' paper out to doctors. Thus all that was received by the latter was a one-page letter urging caution in prescribing. Belatedly in early May the Department decided to send out copies of the paper, but undoubtedly it could have assisted medical practitioners even more by also sending at least a summary of the Elwood/Skegg review. Its actions on the issue, which after all involved public safety, could at best be described as minimal.

No doubt for that reason the PHA called on 8 May for 'clarification of the nature of the relationship between the Department of Health which has to protect the interests of the New Zealand public and the industry that supplies the products that of course are used for the treatment of patients'. Professor Hornblow, PHA President, said: 'It would be important to perhaps see why the Department has taken some time to advise doctors of the risks and whether that process could have been speeded up in some way' (Hornblow, 1989).

Sandra Coney drew parallels with what she described as the 'feeble and mealy mouthed' letters to doctors that the Department had written at the time of the Dalkon Shield controversy and said that the Department's letters were so cautious that doctors would not feel there was anything to be concerned about (Coney, 1989). The question increasingly was whose side was the Health Department on?

The debate continued in the following months when the researchers produced another study on fenoterol. This time they delayed giving it to the Department because they lacked an assurance that the Department would not immediately pass it to the company. The Department then requested the paper from Otago University and promptly passed it on to Boehringer Ingelheim. Again it seems to have bent over backwards to facilitate company comment on research detrimental to its interests while it was under consideration by MARC.

MARC met again on 12 July to consider the second study. It advised me that it did not wish to make any changes to its previous recommendations. The Department advised me on 13 July as follows: 'It is recommended that you accept its advice that no additional action is necessary and that you make a statement reiterating the caution previously issued by the Department'! In fact close scrutiny of MARC's advice showed that it was silent on the question of additional action and that the Department had used some considerable licence in interpreting its recommendation in that way.

It seemed to me that the matter could not rest there. On 19 July I responded saying that the press statement I was issuing would state that the drug would be kept under review, that I understood MARC would be seeking further information, that I expected it to consult its independent epidemiological advisers and the company, and that I also wanted the name of an epidemiologist to be added to the Committee to be submitted to me urgently. The absence of such expertise on issues like this one seemed to me to have serious consequences.

The final act of this saga was played in December 1989. On 8 December the Department advised that it had now received Professor Elwood's further report on fenoterol which it noted I had requested. The report recommended that the use of fenoterol should be minimized. The report was referred to MARC. It met on 6 December and recommended that prescribing of fenoterol be restricted, and that it be withdrawn from the Drug Tariff and available only as a supplementary pharmaceutical benefit thereafter. The Department recommended that I accept the recommendations and ask that PTAC support the recommendations. I agreed. On 18 December the Department advised doctors and pharmacists that fenoterol would be withdrawn from the Drug Tariff and that its further free supply would be available only after application for it as a supplementary pharmaceutical benefit.

Reflections

Looking back over those 21 months as Minister, I feel an enormous sense of frustration that more was not achieved to lower the cost of drugs to the public and to make the State a more effective guardian of the public interest as a drug regulator. Momentum was maintained on a number of fronts but overall, outcomes must be assessed as disappointing.

The power and sophistication of the pharmaceutical industry cannot be underestimated. Those who seek to counter it in the public interest need more than commitment and energy, although that would help. The Health Department will need, in addition, many more skilled operators and continued strong ministerial backing if it is to discharge effectively its public obligations. The PHA workshop's recommendations need to be taken seriously. Every effort has to be made to inform health professionals and the public about the Department's pharmaceuticals strategy, the reasons for it, and the benefits it can bring.

In the absence of strong advocacy in the public interest, the pharmaceutical industry will continue to set the tone of public debate. Its advertising campaign prior to the 1990 general election struck a new low. Not only did full-page advertisements allege that New Zealanders were needlessly suffering because the Government would not pay for new medicines for them, but also its graphic portrayal of a hand with a syringe under the heading 'Should the sick be put out of their misery?' strongly implied that Government policy amounted to euthanasia. Until the public and the professions are better informed about the issues, those impressions will linger at the expense of the public good.

References

Bethell, G.T., President, Pharmaceutical Manufacturers' Association (15 February 1989) Letter to the Prime Minister.

Coney, S. (9 May 1989) Radio New Zealand *Checkpoint*.

Department of Health Draft Press Release (11 April 1989).

Department of Health Press Release (21 April 1989) *Call for Drug Recall Rejected*.

Department of Health (27 April 1989) Memorandum for Minister of Health.

Department of Health (25 June 1990) Memorandum for Minister of Health.

Department of Health (14 July 1989) Memorandum for Minister of Health.

Hardie Marks and Associates (13 July 1990) *Minister Denies Dialysis Patients Drug Treatment* Press Release, issued on behalf of the National Kidney Foundation.

Department of Health (14 July 1990) Memorandum for Minister of Health.

Hardy, Alastair J., Executive Director, RMI (24 February 1989) Letter to Minister of Health.

Holmes, P. (14 November 1989) Television New Zealand *Holmes*.

Hornblow, A. (8 May 1989) Radio New Zealand *Morning Report*.

McLauchlan, W.P., Chief Executive Officer, RMI (6 June 1989) Letter to Minister of Health.

New Zealand Embassy, Washington (10 August 1989) Cable to Minister of External Relations and Trade.

Pharmacology and Therapeutics Advisory Committee (PTAC) (2 June 1989) A Report to the Minister of Health.

Public Health Association of New Zealand (PHA) (1990) *Pharmaceutical Advertising and Promotion — Options for Action* Wellington, Public Health Association of New Zealand.

Researched Medicines Industry (RMI) (20 April 1989) *Medicines Industry Calls for Withdrawal of Drugs* Press Release.

Researched Medicines Industry (RMI) (1990) Minutes.

Salmond, G., Director General of Health (6 October 1988) Press Release.

Adverse Reactions: The Fenoterol Saga

Neil Pearce

In April 1988 I was approached by three colleagues in the Wellington School of Medicine (Julian Crane, Richard Beasley, and Carl Burgess) who asked me to become involved in their research on the safety of fenoterol, a beta agonist asthma drug which they believed could be responsible for an epidemic of asthma deaths in New Zealand. Three years later, we had conducted three major epidemiological studies and a dozen laboratory studies, the availability of the drug had been severely restricted in New Zealand and Australia, the company which manufactures the drug (Boehringer Ingelheim) had agreed to reduce its dose in other countries, and an editorial had appeared in the prestigious international medical journal the *Lancet* calling for the use of beta agonist drugs in asthma to be redefined. In the interim, our work had undergone three major Health Department reviews, and at least three major company reviews, the company had apparently threatened the Health Department with legal action on at least two occasions, two different issues had been referred to the office of the Ombudsman, and we had met with our own lawyer on more than 10 occasions.

The fenoterol saga is very long and complex, both in the scientific sense and in the political sense. In some ways, it is a rather extreme example of the problems of pharmaceutical promotion in New Zealand, but much can often be learnt from studying extreme examples. I do not intend to cover all of the issues here, but rather to concentrate on those issues which are most relevant to pharmaceutical promotion in New Zealand. However, even in this more limited context, it is important to first discuss some of the background to the fenoterol debate.

Some History

The 1960s asthma death epidemics

The story begins in the 1960s when asthma deaths increased suddenly in several countries, including the United Kingdom, Australia, and New Zealand. These sudden epidemics of asthma deaths were very surprising, since asthma death rates had been relatively low for the previous 100 years

(Speizer and Doll, 1968). It was found that the epidemics were not due to changes in diagnostic criteria, death certification, or coding practices, or due to changes in the prevalence of asthma or in environmental factors which may trigger asthma attacks. It was therefore concluded that the epidemics must be due to an increase in case-fatality due to new methods of treatment for asthma. In particular, it was noted that the epidemics followed the introduction of pressurized beta agonist inhalers (also known as 'aerosols', 'puffers' or 'pumps') (Speizer et al., 1968). These were very popular as they are very effective at relieving asthma symptoms. Sales rocketed in many countries, and the epidemics of deaths generally paralleled the increase in sales. It was also noted that the increase in deaths was greatest in the 10–19 year age-group and that 'at these ages children have begun to act independently and may be particularly prone to misuse a self-administered form of treatment' (Speizer et al., 1968).

Possible mechanisms

There were three main concerns about the potential hazards of beta agonists, and particularly about isoprenaline which was the most commonly sold drug (Pearce et al., 1991). These had been raised when this drug first became available in the 1940s and 1950s, but the concerns were greater when it was marketed in pressurized inhalers in the 1960s.

The first concern was that the regular use of beta agonist inhalers, while providing symptomatic relief, could result in worsening asthma (Pearce et al., 1991). These chronic effects were demonstrated in several studies of isoprenaline.

A second concern was that the use of inhaled beta agonists in an asthma attack, by providing symptomatic relief, could lead to delays in seeking medical help (Pearce et al., 1991). Almost all asthma deaths occur outside hospital, and in many instances death could have been avoided if medical help had been sought at an earlier stage of the attack.

The third concern was the potential for toxicity when inhalers are used in an asthma attack, since beta agonists, and particularly isoprenaline, have strong adverse effects on the heart (Pearce et al., 1991). Studies in animals have found that these side-effects are not particularly dangerous in normal circumstances, but that quite small doses of isoprenaline can cause death when the animal is suffering from oxygen deprivation (hypoxaemia), a condition which commonly occurs in severe asthma attacks.

The isoprenaline forte hypothesis

Despite this evidence, the isoprenaline hypothesis was regarded with scepticism by some clinicians who were reluctant to consider that their treatment could have caused deaths. This scepticism was encouraged by the fact that there had been large sales of inhalers in the United States and several other countries which did not have mortality epidemics. This confusing situation was clarified by Paul Stolley who noted that a high dose version of isoprenaline (isoprenaline forte) had only been licensed in eight countries; six of these (England and Wales, Ireland, Scotland, Australia, New Zealand, and Norway) had mortality epidemics which coincided with the introduction of the drug, and in the other two countries (the Netherlands and Belgium) the forte preparation was introduced relatively late and sales volumes were low (Stolley, 1972). No mortality epidemics occurred in countries such as Canada and the United States in which isoprenaline forte was not licensed. Consequently, an editorial in the *British Medical Journal* which reviewed Stolley's work was entitled 'Asthma deaths: a question answered' (Editorial, 1972) and death from beta agonist inhalers was designated one of the most important adverse drug reactions since thalidomide (Venning, 1983).

History is Rewritten

By the early 1980s, the prevailing wisdom had changed, and the isoprenaline forte hypothesis was rejected in many medical texts and reviews (e.g. Olson, 1988). However, no significant new evidence had appeared in the interim, and the process of 'reinterpretation' of the 1960s epidemic was based on minor anomalies in the time trend data, which were emphasized, and to some extent exaggerated, in subsequent reviews (Pearce et al., 1991). In addition, other factors such as delays in seeking medical help began to receive greater emphasis, even though it was very difficult to envisage how such delays could have spontaneously caused the 1960s mortality epidemics unless they were themselves caused by the introduction of a new drug.

Orciprenaline and the Food and Drug Administration

The tendency to discount the role of beta agonists received increasing emphasis as a result of the controversy following the decision of the United States Food and Drug Administration (FDA) in 1983 to license orciprenaline (a beta agonist drug which has similar cardiac side-effects to those of isoprenaline) for non-prescription sale. The FDA decision was rescinded two months later, because of the concerns regarding the role of inhaled beta agonists in the 1960s mortality epidemic (Hendeles and Weinberger, 1983). Two reviews which were subsequently published in leading medical journals questioned the role of inhaled beta agonists in the 1960s mortality epidemic; one (Lanes and Walker, 1987) specifically acknowledged funding from the manufacturers of orciprenaline (Boehringer Ingelheim), whereas the other review (Esdaile et al., 1987), which cited the FDA decision on orciprenaline as its justification, came from a group which has disputed most of the major epidemiological discoveries in the past 15 years, including such well-established associations as: tobacco and lung cancer; oral contraceptives and stroke; diethylstilbestrol and vaginal cancer; oestrogens and endometrial cancer; tampon use and toxic shock syndrome; and aspirin and Reye's syndrome (Savitz et al., 1990). Both reviews dismissed the striking time trends in the six countries where isoprenaline forte was heavily sold, and instead emphasized the minor anomalies in the time trend data for other countries. They presented no alternative hypotheses to explain the mortality epidemics, other than to propose that they may have been due to deaths from respiratory infections being misclassified as deaths from asthma, or that there may have been an increase in asthma incidence. These hypotheses had already been rejected by investigators who had previously studied the epidemic. However, despite the obvious shortcomings of these two reviews, and the circumstances in which they were written, they have subsequently been extensively quoted, and until recently many respiratory physicians believed that the isoprenaline forte hypothesis had been refuted (Pearce et al., 1991).

Déjà Vu

In 1976, a second asthma mortality epidemic began in New Zealand, but not in other countries. It was first reported in 1981 by a Professor of Immunology at the Auckland Medical School, Doug Wilson (Pearce et al., 1991), who noted an increase in young people dying suddenly from acute asthma in Auckland. He proposed that the sudden nature of the deaths suggested a heart attack, possibly due to overuse of asthma drugs. Rod Jackson (who is the co-author of another chapter in this book) studied the second New Zealand epidemic in depth, and concluded that it appeared to be real, and could not be explained by changes in the classification of asthma deaths, inaccuracies in death certification, or changes in diagnostic fashions. He concluded that it was very unlikely that the epidemic could be due to changes in the incidence or prevalence of asthma in New Zealand, and that the most likely explanation, as for the 1960s epidemics, appeared to be an increased case-fatality rate related to changes in the treatment of asthma in New Zealand (Jackson et al., 1982).

The New Zealand asthma mortality survey

A national asthma mortality survey which was subsequently undertaken by the Asthma Task Force included all of the 271 asthma deaths in persons under the age of 70 years during mid 1981 to mid 1983 (Sears et al., 1985). However, the mortality survey did not include a control group, and (with one minor exception) the findings for individual asthma drugs were not reported. This reluctance to consider the possibility that the epidemic might be iatrogenic was surprising given the previous conclusions of Rod Jackson (Jackson et al., 1982). It may have been partly due to the difficulties of investigating this hypothesis, but it may also have reflected the prevailing sentiment amongst respiratory physicians that asthma drugs were not a major cause of asthma deaths.

The first report from the mortality survey did not specify any criteria for overuse, but concluded that 'excessive use of bronchodilator drugs did not account for the high mortality rates' (Sears et al., 1985). In contrast, subsequent reports by the same authors noted that 'most patients whose final episodes had lasted for several hours had repeatedly used their inhaled bronchodilator aerosol . . . to administer large doses of beta-agonist, without seeking additional therapy' (Sears and Rea, 1987). Nevertheless, the initial dismissal of a possible role of beta agonist inhalers has since been widely quoted, and until recently most New Zealand respiratory physicians apparently believed that the mortality epidemic was not iatrogenic, even though no other plausible explanation for the epidemic had been suggested (Pearce et al., 1991).

The Fenoterol Studies

Thus, there was no tenable explanation for the second New Zealand mortality epidemic until the fenoterol hypothesis was developed by our group. This hypothesis stemmed initially from published reports of the greater cardiovascular side-effects of fenoterol, and from incidental information in two previously published studies indicating that fenoterol was used by a relatively high proportion of asthmatics who died (Pearce et al., 1991). Fenoterol, a beta agonist which was marketed in a high-dose preparation, was introduced to New Zealand in April 1976, and the epidemic of deaths began in the same year. There was a rapid increase in the fenoterol market share to about 30 per cent, and a similar rapid increase in the death rate, between 1976 and 1979. In contrast, the drug represented less than 5 per cent of the market in most other countries and was not available in the United States. Although the market share in West Germany was 50 per cent, use of inhaled beta agonists was relatively low in that country, and the *per capita* usage in New Zealand was greater than that of any other country, and more than three times that in West Germany (Pearce et al., 1991). The reasons for the high sales of fenoterol in New Zealand are unclear but are probably related to an effective marketing programme. In particular the company established good contacts with prominent New Zealand respiratory physicians through a series of annual asthma symposia in Rotorua.

Experimental studies of fenoterol

Following these observations, our group conducted an experimental study which found that repeated use of fenoterol resulted in side-effects which were greater than those of other commonly used beta agonists, and were even greater than those of isoprenaline. These findings were similar to those in three previous studies which had tested fenoterol in relatively large doses, and have also been confirmed in subsequent studies conducted by our group and by others (Pearce et al., 1991). The experience with previous animal studies of isoprenaline suggests that although these side-effects are not dangerous under normal conditions, they can be dangerous under conditions of oxygen deprivation in a severe asthma attack.

The first case-control study

Thus, by 1988 the epidemiological and experimental evidence strongly suggested that the high use of inhaled fenoterol could be the main cause of the second New Zealand asthma mortality epidemic. This evidence was comparable to that which had linked isoprenaline forte to the 1960s mortality epidemics. It had been generally agreed in discussions of the 1960s epidemic that the definitive method for testing such hypotheses was to conduct case-control studies of asthma deaths. The case-control design, in the context of studying asthma deaths, involves comparing persons who died of asthma (cases) with a sample of patients with non-fatal asthma (controls) in order to determine what factors are associated with an increased risk of death (Pearce et al., 1990). An important issue for a study of whether a particular asthma drug is associated with an increased risk of asthma death is the need to control for other risk factors for asthma death, such as chronic asthma severity. Thus, the ideal approach is to compare patients who died of asthma (cases) with a group of patients with non-fatal asthma of a similar chronic severity (controls).

A suitable study design had been suggested in an internal Boehringer Ingelheim memorandum of October 1988 which stated that:

By comparing the exact drug medication of patients who died with hospital controls matched on the basis of severity (of past hospital admissions, of recent steroid use, and other factors) then any major disparity between the representation of [fenoterol], or other drugs, and the fact that they were compared with matched hospital controls, could be taken as positive evidence, and strong positive evidence, of a possible selective toxicity of the over-represented agent.

This approach was consistent with the findings of the only previously published case-control study of asthma deaths (Rea et al., 1986). Accordingly, our group conducted a case-control study to investigate the possible role of fenoterol in the second New Zealand asthma mortality epidemic, using a design very similar to that outlined in the Boehringer Ingelheim memorandum (Crane et al., 1989). The study was declined for funding by the Medical Research Council and the Asthma Foundation, but our group managed to carry out the study in our spare time at minimal cost. (These two organizations have subsequently been very supportive of our research, and in mentioning these historical difficulties I do not intend any criticism of them in their present form.)

The study was partly based on data which had been collected for the previous New Zealand asthma mortality survey, but not previously reported; the cases (deaths) comprised all asthma deaths in New Zealand in the 5–45 year age-group during 1981 to 1983, and for each case four controls were selected from records of hospital admissions for asthma during the same period. Information on prescribed drug therapy at the time of the last attack was documented for cases and controls. The only asthma drug which was found to be significantly associated with asthma deaths was fenoterol; in patients with very severe asthma, the death rate in those prescribed fenoterol was more than 10 times that in those prescribed other beta agonist inhalers (Crane et al., 1989).

Round One

Prior to submission for publication

The case-control findings were very striking, and very consistent with the other epidemiological and experimental evidence. Nevertheless, we were aware that the findings of the case-control study were likely to come under attack, and this criticism began before the study had even been submitted for publication. The Department of Health therefore convened an independent Review Panel to examine the research prior to its submission for publication. The Review Panel met with our research group and with members of the Asthma Task Force on 20–21 December 1988. Although members of the Review Panel had some criticisms of the study design, they concluded that: 'They would accept the study design as appropriate to the study and that . . . the findings of the study would be sufficient to justify public health action' (Leeder, 1988).

However, the Asthma Task Force questioned the accuracy of the data used, and it was therefore agreed that the paper would be delayed for a further six weeks so that the data for all of the deaths and a sample of the controls could be re-checked jointly by our research group and by the Asthma Task Force. This process showed that, although there were still disagreements about the general study design, there was very good agreement on the classification of the key medication data. The report by one Task Force member stated that 'these are the best data that will ever be available to try to answer this extremely important question'. The few discrepancies were resolved by incorporating the changes suggested by the Asthma Task Force.

Prior to publication

The paper was submitted to the *Lancet* on 3 February 1989. The Journal referred the paper to two independent referees who cleared it with very minor alterations to the text, and it was accepted for publication on 20 February 1989.

When the paper was submitted for publication, we gave a copy to the Department of Health. Our group had repeatedly expressed our concern to the Department of Health that there might be interference in the normal process of scientific peer review and publication if the manuscript were given to the manufacturer of fenoterol (Boehringer Ingelheim). However, we finally agreed for the then Director-General of Health to give a copy to the company on the written condition that the company would make no attempt to interfere with the paper's publication. Accordingly, on 7 February 1989 a copy of the manuscript was sent to the Medical Director of Boehringer Ingelheim (New Zealand), Doug Wilson (the former Professor of Immunology who had first raised the possibility that asthma drugs could have caused the second New Zealand epidemic).

The first indication that all was not well came when I was visiting a very eminent epidemiologist in the United Kingdom. He had reviewed an earlier draft of the paper for our group, and consequently (with his permission) had been acknowledged in the published version of the paper. As a result, the company had sent him a copy of its internal review of the manuscript (dated 8 February 1991). The document commenced with the statement that:

There may be no formal protocol for the study. Information on this defect was provided by two Task Force members . . . It appears that the protocol as such was developed during the course of the study as information came to hand.

This statement was, of course, incorrect and had not been raised with our group by Boehringer Ingelheim or the Asthma Task Force. The influence of such incorrect statements on reviewers of the paper is a matter of speculation, although it is notable that at least one Boehringer Ingelheim reviewer subsequently referred to the possible lack of a formal protocol. The Boehringer Ingelheim document also argued that: 'This study should be attacked at its roots, i.e. the study design'. Soon afterwards, the first reviews commissioned by Boehringer Ingelheim began to arrive at the Department of Health, for consideration by the Medicines Adverse Reactions Committee (MARC). Two of the first reviews to be received were written by the same groups which had previously disputed the role of isoprenaline forte in the 1960s epidemics, as a result of the FDA decision on the availability of orciprenaline (another Boehringer Ingelheim asthma drug). One review (from A.R. Feinstein) stated that:

I shall be very surprised if the work is accepted by The Lancet . . . The editors of The Lancet, however, sometimes do not seek a careful external scientific peer review; and many editors nowadays are attracted by potential publicity for published reports. If the work is accepted and published, the devastating criticism and subsequent embarrassment (for authors, editors, etc) will occur afterward.

The other review (from S.F. Lanes and A.W. Walker) gave more substantive criticisms and outlined the approach that had been followed:

Our approach was to review the study, identify potentially important sources of bias . . . The next step would be to evaluate the importance of each source of error . . . Exposure classification is a concern because subjects were classified according to drugs that were prescribed but not necessarily used . . . Controls were matched to cases by region; however, this variable was not considered a potential confounding variable.

What this review neglected to mention, however, was that the above two 'sources of error', as with many other problems which were included in a long list, would tend (and in some cases are guaranteed) to produce false negative (rather than false positive) results. In other words, they were very unlikely to account for the positive findings of the study. Another feature of the Boehringer reviews was their lack of consistency. For example, one review (from S. Buist) stated that: 'The first and probably most important problem [with the study] is the absence of an initial hypothesis' whereas another review (from A. Reebuck) stated that: 'The authors [of the study] have developed a hypothesis which they have pursued relentlessly'.

A subsequent 'consensus meeting' of Boehringer Ingelheim reviewers, held in New York on 1–2 April 1989, concluded that: 'The study design is seriously flawed and may lead to unjustified policy formulation and prescribing decisions' (Buist et al., 1989).

However, a contrary view was subsequently provided by another group of reviewers who stated that: 'Like earlier correspondents some of us were hired by Boehringer-Ingelheim to review a draft of the fenoterol paper . . . Unlike those correspondents, however, we did not attend the meeting convened by the company — and we have reached different conclusions' (Sackett et al., 1990).

Although some reviews were relatively balanced and made some positive comments about the study, the overall weight of the Boehringer Ingelheim reviews was overwhelmingly negative, and the substantial evidence against fenoterol was essentially dismissed. This contrasted markedly with the conclusions of independent reviewers, including those commissioned by the Department of Health (Elwood and Skegg, 1989).

The approach of the Boehringer Ingelheim reviewers was not surprising, and was to some extent predictable, given the history of similar epidemiological controversies. What was surprising in this instance was that some authorities were apparently influenced by these reviews, and even gave them equal weight with their own independent reviews. In particular, these reviews were sent to the *Lancet* by Boehringer Ingelheim, and on 3 March 1989 the *Lancet* wrote to our research group rescinding its previous unconditional acceptance of the manuscript for publication stating that:

Some of the [Boehringer Ingelheim] commentaries have now been sent to us and they are causing some anxiety . . . Three possible courses of action face a journal having second thoughts about a paper accepted but not yet published . . . withdrawal . . . counter convincingly the very critical comments of the Task Force . . . publication in the second half of the journal accompanied by a highly critical editorial in the same issue.

In our reply of 17 March 1989, our research group stated that:

We are concerned that our scientific paper, which was accepted unconditionally . . . could have this unconditional acceptance withdrawn following submissions made by a pharmaceutical company . . . we can counter convincingly the very critical comments . . . we look forward to seeing [the paper] in print.

The *Lancet* accepted these arguments, and the publication date was set for 29 April 1989.

In the interim, the Boehringer Ingelheim lobbying was also operating in other spheres. On 15 March 1989, the New Zealand Director-General of Health wrote an internal memorandum stating that:

Doug Wilson left me with the attached press releases he and Boehringer have prepared for release in the event of a media expose. Doug is keen to work in with us over press releases and other contact with the media and using the opportunity for some good asthma education. In that regard I think Doug's background is excellent.

However, this joint approach was apparently not proceeded with, and just prior to the publication of the paper, the Department of Health and Boehringer Ingelheim clashed over the Department's draft media release and its plan to send an abstract of the *Lancet* paper to all New Zealand doctors. Boehringer Ingelheim wrote to the Department that:

We are most concerned at your intended communication with doctors . . . Enclosing a copy of the abstract suggests Departmental approval . . . unless the Department's approach is modified we are left with no option but to take every step available to us to protect ourselves.

With regard to the Department's letter to doctors, Boehringer Ingelheim stated that:

We have learned that the content and ambiguity in your letter has caused substantial anxiety and confusion among asthma patients and doctors . . . the responsibility for the medical and legal consequences rest with you . . . We expect [your next] letter to state that therapeutic conclusions cannot be drawn from the Crane et al. study.

The Director-General of Health subsequently advised the Department not to circulate the abstract of the *Lancet* paper, but this decision was effectively reversed in the following week when a complete copy of the paper was sent to all doctors in New Zealand.

The Department of Health's response to the company's lobbying was not helped by the fact that MARC did not include an epidemiologist. However, the Department had commissioned its own review of the study by epidemiology Professors Mark Elwood and David Skegg. This, like the other independent reviews, was quite different in style and conclusions from the reviews commissioned by the company. It contained some criticisms of the study, but concluded:

We recommend that action is taken on the basis that this study suggests there may be an increased risk of death from asthma in patients prescribed fenoterol by metered dose inhaler who have severe disease . . . This evidence, although far from conclusive, suggests that the current practice in the use of feneterol should be modified (Elwood and Skegg, 1989).

After publication

The paper was finally published in the *Lancet* on 29 April 1989, and on that date the Health Department sent out a one-page letter advising doctors that: 'Fenoterol will not be withdrawn from the market, but doctors should review and perhaps modify the treatment of severe asthmatics'.

However, this cautious advice was pre-empted by press releases from the company and the Asthma Task Force on 28 April. The company's publicity package (the first of many over the ensuing two years) was couriered to most doctors, pharmacists and health reporters in New Zealand, and was apparently also delivered to many respiratory physicians and pharmacoepidemiologists in other countries. It highlighted the conclusions of the New York meeting of Boehringer Ingelheim reviewers that the study had 'serious flaws in design, execution and analysis which rendered its results uninterpretable'. These conclusions received extensive coverage in the New Zealand media with headlines such as 'Asthma study flawed say foreign experts'.

The Boehringer Ingelheim package also contained other material which claimed to demonstrate a lack of association between fenoterol sales and asthma deaths. For example, a graph was presented of fenoterol and sales, and asthma deaths in New Zealand, which showed fenoterol sales rapidly rising during the 1980s while the death rate rapidly declined; however, one crucial point on the graph was incorrect and another point was only provisional; and a corrected graph showed a strong correlation between fenoterol market share and asthma deaths (figure 1, page 87).

A second graph which claimed to show no correlation between fenoterol sales and asthma deaths in West Germany was drawn on an inappropriate scale, and a corrected graph once again showed a correlation between fenoterol sales and asthma deaths (figure 2, page 88). However, attempts to correct these inaccuracies were lost in the publicity about the condemnation of the study by 'foreign experts'. Furthermore, the issues under dispute were highly complex, and were poorly understood by many New Zealand doctors and health authorities, most of whom have little or no training in epidemiology. At best they were left with the impression that the experts could not agree; at worst they were persuaded that the study was 'fatally flawed'.

Round Two

The second case-control study

The majority of the criticisms which were levelled at the first case-control study by Boehringer Ingelheim reviewers involved problems which

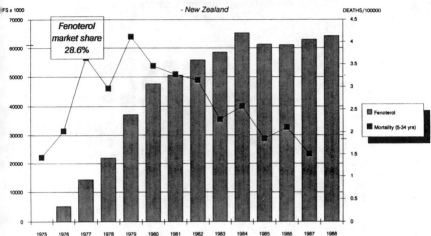

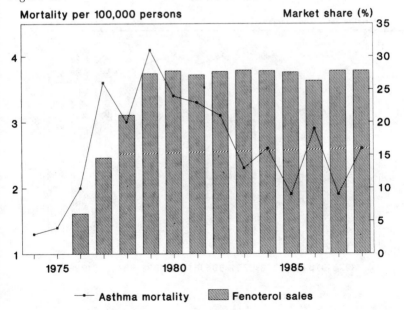

Figure 1: Comparison of the Boehringer Ingelheim graph and a corrected version: New Zealand.

Figure 2A

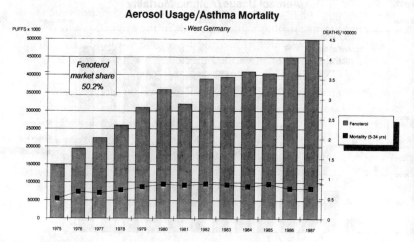

Figure 2B

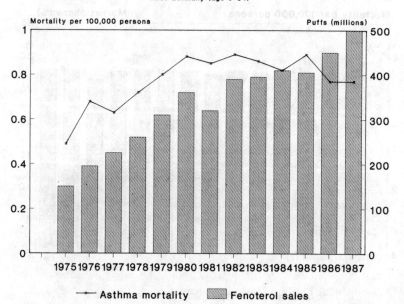

Figure 2: Comparison of the Beohringer Ingelheim graph and a corrected version: West Germany.

were very unlikely to occur, were trivial, would tend to produce false negative results, or were simply incorrect. The main criticism which had some validity was one that had been noted in the original report (Crane et al., 1989), and was also stressed in the review of Elwood and Skegg, namely that the data for prescribed medicines was taken from different sources for the cases and controls (Elwood and Skegg, 1989). Our research group was already aware of this criticism, and had developed a new study design which enabled the information on prescribed drug therapy to be collected from the same sources for cases and controls. Using this design, we conducted a second New Zealand case-control study of asthma deaths in 5 to 45-year-olds during 1977 to 1981 using this new design (Pearce et al., 1990). This confirmed the findings of the first study that fenoterol was associated with asthma deaths, and that in patients with very severe asthma the death rate in those prescribed fenoterol was much higher than in those prescribed other beta agonist inhalers.

The response

This second study was presented on 29 June 1989 at a Pharmacoepidemiology Symposium in Newcastle, Australia. The symposium, which included pharmacoepidemiologists and regulatory authorities from Australia, New Zealand, the United Kingdom, and the United States, concluded that the new study overcame the principal problems with the previous study, and that the evidence was now sufficiently strong that the drug should be withdrawn in New Zealand. One of the principal speakers, Professor David Lawson, Chairman of the United Kingdom Committee for Review of Medicines, was subsequently contacted by the New Zealand Department of Health and replied that:

These findings are very difficult to explain short of accepting the probability that fenoterol is causally associated in some way with asthma deaths . . . I said at the conference, that were I was to be advising the Minister of Health in New Zealand I would indicate that, in my judgement, the new data provides sufficiently strong evidence to justify removing fenoterol from the market place in the very near future . . . other bronchodilators are available and its loss would not constitute a public health hazard.

The response to this new study posed a problem for our research group in that it meant that the manuscript should be made available to the Department of Health. However, because of the problems which had occurred with the publication of the first study, we were anxious that the company should not be given a copy of the second paper until it had been accepted for publication. Although the Australian Department of Health could not give an absolute guarantee on this matter, it did give reasonable assurances, and a copy of the manuscript was made available to it immediately. However, the New Zealand Department of Health declined to give any assurances on this issue; instead it sought a copy of the manuscript from the University of Otago (our employers) using the Official Information Act. We were obliged to hand over the manuscript, and the Department of Health then passed it on to the company. In fact, although there was apparently no direct interference with the publication of the second paper, one group of Boehringer Ingelheim reviewers did attempt to publish a critical letter in the *New Zealand Medical Journal* prior to the publication of the paper in *Thorax* (a leading respiratory disease journal published by the British Medical Association).

MARC (which still did not include an epidemiologist) then met to consider the new study. The minutes of the meeting (which were released by the then Minister of Health, Helen Clark, who is the author of another chapter in this book) show that the committee agreed that the study implied that fenoterol was associated with an increased risk of death, but advised the Minister that no further action on fenoterol was required. Instead it was suggested that animal studies should be conducted in Dunedin to ascertain the mechanism of death.

During the same period, the company was attempting to obtain Medical Research Council (MRC) approval for research designed and funded by the company. The MRC asked for comment from 34 experts and received 16 replies within three weeks. The MRC reported that it had received complimentary comments from only three people, two being members of the Asthma Task Force. The MRC decided to withdraw from the research when it was revealed that Boehringer Ingelheim was arranging for direct funding of the research which would be conducted by a member of the Asthma Task Force. The Director of the MRC stated that this Task Force member had been 'attempting to set up an independent mechanism which would bypass council assessment . . . I would wish to question the appropriateness of establishing two separate mechanisms for the funding of substantially similar research projects'. The MRC invited Boehringer Ingleheim to instead make 'unrestricted funds' available for asthma research, but the company has apparently not accepted this invitation.

The Boehringer Ingleheim reviewers prepared their response to the second study at a meeting at the Beverly Hills Wilshire Hotel on 28 July 1989. Our research group was excluded from the meeting, but the Director of the MRC attended the meeting as an observer. His report was complimentary about some aspects of the proceedings, but noted that:

[The participants] were under some pressure [during the preparation of the final report] from the PR representative . . . whose task was ostensibly to translate a technical report into lay language but who clearly held partisan views in favour of Boehringer Ingelheim . . . The company representatives prepared the ground for the meeting but stood back from the actual proceedings . . . Nevertheless their presence could not be ignored and their hospitality was generous, to say the least. The effect of this process on scientific objectivity is always a matter of speculation (Hodge, 1989).

The consensus meeting concluded that: 'The second study avoids only one of the methodologic problems of the first study . . . but has retained others and introduced new methodologic problems' (Buist et al., 1989).

The Department of Health also commissioned an independent review of the second study from Professor Mark Elwood, who reached the very different conclusion that:

The balance of the available information is in favour of the causal rather than the confounding hypothesis . . . It is recommended that the drug regulatory authorities should take steps to ensure that the use of fenoterol is minimised (Elwood, 1989).

The controversy in New Zealand was effectively resolved in December 1989 when it was announced that the drug was to be removed from the Drug Tariff. This action followed a recommendation from MARC (which by now included an epidemiologist) as a result of its examination of the Elwood review. Similar action was announced in Australia in March 1990, and the company subsequently reduced the dose of the drug in the United Kingdom and other countries.

Epilogue

More recently, our group conducted a third national case control study (funded by the Asthma Foundation) of deaths during 1981 to 1987 (Grainger et al., 1991); two control groups were used in order to address the remaining criticisms which had been made of the previous studies. Whichever control group was used, fenoterol was associated with an increased risk of asthma death, but the alternative control group suggested by Boehringer Ingelheim critics of the previous studies actually yielded stronger relative risks than the approach used previously.

Further evidence that the association between fenoterol and asthma death is likely to be causal was provided by two studies recently published in the *Lancet*. In particular, Sears et al. found that the regular use of inhaled fenoterol was associated with greater morbidity in patients with mild to moderate asthma (Sears et al., 1990).

The strongest evidence in support of the New Zealand findings has recently come from a Boehringer Ingelheim-funded study conducted in Canada by several members of the Boehringer Ingelheim consensus panels (Spitzer et al., in press). This study examined 44 asthma deaths in the province of Saskatchewan during 1980 to 1987, and 233 controls. Nearly half of the deaths had been prescribed fenoterol; overall the death rate in those prescribed fenoterol was five times that in those prescribed other beta agonists. However, these results (which confirmed the New Zealand findings) were not directly mentioned in an abstract which was circulated widely prior to the publication of the paper (Bown, 1991). Instead, the abstract focused on the possibility of a general beta agonist effect because the study had raised the possibility that there could also be problems with other beta agonists at high doses. The fact that the risk of death was higher with fenoterol than with other beta agonists was only mentioned briefly, and was then dismissed on the grounds that this was merely a dose effect and that the risk would have been less if fenoterol had been marketed in a lower dose (this assertion is contradicted by laboratory studies which show greater cardiac side-effects with fenoterol even at lower doses, and by epidemiological evidence that asthma mortality epidemics have only occurred with isoprenaline forte and fenoterol and not with other beta agonists). This abstract received extensive publicity in the *New York Times*, the *Lancet*, and the *New Scientist* (Bown, 1991) prior to the publication of the full paper (Spitzer et al., in press), thus leading to confusion about the safety of beta agonists in general, and obscuring the fact that the study had confirmed the New Zealand findings regarding fenoterol.

Finally, the most important confirmation of the fenoterol hypothesis came with the news that the asthma death rate in New Zealand had dropped markedly following the publication of our first study.

Issues for the Future

There is an old saying that those who do not learn the lessons of history are condemned to repeat it. It is tempting to speculate whether the second New Zealand mortality epidemic could have been avoided if knowledge gained from studies of the 1960s epidemics had been properly heeded. Certainly, the greater potency and greater cardiac side-effects of fenoterol were known before its introduction and widespread use, as was the fact that fenoterol could cause death in baboons when infused in doses previously given to humans. Thus it is important that the lessons of the isoprenaline and fenoterol debates are not ignored for a second time. More generally, it is important that new drugs are tested under the conditions and in the doses in which they are likely to be used, and that evidence of side-effects is considered seriously.

As noted above, the fenoterol story provides a rather extreme example of the problems of pharmaceutical promotion in New Zealand. I do not intend to fully discuss all of the relevant issues here, but rather to raise some general issues, some of which are addressed elsewhere in this book. In doing so, I do not intend to imply that all, or indeed any, of these issues relate specifically to the fenoterol saga.

The company and its consultants

A recent article about the fenoterol controversy in the *Epidemiology Monitor* raised the issue as to what limits should be set on a company's efforts to defend its products which have been implicated in serious side-effects (Bernier, 1990). Clearly a company has a right to argue against what it believes is weak data or incorrect conclusions, and it is not in the interests of society or the company to withdraw a drug which has been incorrectly implicated. However, a company's right to defend itself does not extend to interference with publication or impugning the integrity of the researchers concerned. Moreover, a company also clearly has a moral obligation to seek the truth of the matter when seeking advice from consultants, rather than just to prepare the 'case for the defence'. This is not to imply that deliberate corruption occurs; in fact this appears to be very rare. However, a company which intends to prepare the 'case for the defence' may seek out academics who (usually because of sincerely held beliefs) have been very critical of similar studies in the past. Thus, the shaping of the 'case for the defence' usually involves 'selection' rather than 'coercion' of experts, although subtle forms of influence may also occur.

The response of consultants may also heavily depend on what question is posed by the company. One possible question would be 'Is there any chance that the data are right?', to which the answer is invariably 'Yes'. Most commonly, however, the question which is posed is 'Is there any chance that the data are wrong?', to which the answer is also invariably 'Yes'. This is perhaps the question which is most appropriate in the scientific context, where the emphasis is on scientific criticism and debate. However, in the context of public health decision-making, the most appropriate question is 'On balance, what conclusion is most likely true from the data?'. Clearly, quite different reviews will eventuate depending on which question is asked.

A related issue is the attitude of the consultants to the reviewing process. Although criticism is an important component of the scientific process, an overemphasis on criticism (sometimes justified by crude interpretations of the Popperian philosophy of science) can lead to almost any scientific study being dismissed as 'fatally flawed'. As Stolley writes:

> A . . . distressing development has been the attitude of some self-proclaimed pharmacoepidemiologists that their job is to attack competent studies as consultants to drug companies, who pay them handsomely and even award grants to their research unit as a form of reward. Often the sponsorship of these 'disinterested reviews' is not clearly stated. The most pernicious of these articles are characterized by an unwillingness to focus on the totality of the evidence and a concentration on real and imagined flaws that could not possibly account for strong associations (Stolley, 1989).

This process can produce an apparent consensus which is quite different from the real consensus of independent scientists. This can be very influential since, although science is based on criticism and debate, public health policy is generally based on consensus. Thus, the selection by the company of a few scientists who are hypercritical of others' work, can result in massive pressure on public health decision-makers. This pressure is particularly effective since it apparently comes from independent scientists, whereas it would not be taken so seriously if it came directly from the company. In this sense, the company's reviewers have the privilege of acting as 'lawyers for the defence' while maintaining the image of being an independent jury.

The journal
This type of reviewing also poses problems for a journal which is considering for publication a paper which implicates a particular drug in serious side-effects. Ideally, a good journal should stand by its own reviewing process, and should not consider any unsolicited reviews. However, as the *Epidemiology Monitor* points out, if reviews commissioned by a company are sent to a journal: 'the journal is placed in a difficult position because it may feel that it cannot prudently ignore the criticisms, yet they may not have been obtained during the normal process of peer review' (Bernier, 1990).

The researchers
Such reviews also pose a dilemma for researchers who have discovered evidence that a particular drug may have serious side-effects. In theory, any evidence of hazard should be made immediately available to the scientific community, and should have some influence on public health decision-making. In practice, researchers who have discovered evidence that a particular drug may be hazardous require very strong evidence,

a fair amount of perseverance, a sense of humour, and a good lawyer. Even then, there is the danger that, despite the best of intentions, researchers may over-react to the resulting wave of criticism and may tend to over-state the case against the drug, particularly if they consider that the company's criticisms are trivial, irrelevant, or incorrect. As Stolley writes:

The pharmacoepidemiologist must develop a thick skin; this is not a field for timid souls. More important, the pharmacoepidemiologist must have some historical and ideologic anchorage and perspective to be able to understand the nature of the attacks. The history of drug regulation, the battle for effective drug efficacy and safety standards, and the connection of drug regulation with the sanitary movement of public health are all a part of the progressive tradition we have inherited as epidemiologists (Stolley, 1989).

Ultimately, the best approach for researchers is to address any criticisms in further studies; certainly, this approach was the key factor which eventually led to the resolution of the fenoterol saga.

The Health Department

Given the unfortunately combative environment in which debates on drug safety occur, it is important that the Department of Health should have advice from independent epidemiologists and pharmacologists who are not connected with the researchers or with the company. In particular, most reports of adverse drug reactions involve epidemiological data, and committees such as MARC should therefore include several epidemiologists.

There are two key questions for the Department of Health: 'Under what conditions should a drug be licensed?'; and 'Under what conditions should this licence be restricted or revoked?'. The fact that fenoterol was not licensed in the United States might be taken as sufficient reason for its licence to be declined in New Zealand; on the other hand the drug was licensed in most other Western countries and it would perhaps be harsh to blame the New Zealand Department of Health for making the same mistake.

It is reasonable to assume that fenoterol would never have been licensed in any country if the information which is available now had been available at the time of application for licensing (or even if the evidence which was available then had been properly and comprehensively considered). However, it can be extraordinarily difficult to withdraw or restrict a drug once it has been licensed. No Department of Health will wish to place itself at risk of legal action by taking action against a particular drug unless the evidence of hazard is very strong. Furthermore, the public is not well served if a Department takes premature action on the basis of false alarms. In fact, it might be argued that most 'scares' about the safety of drugs, or other chemicals, turn out to be false alarms. However, the danger is that the frequency of such false alarms may create a sense of apathy, and that an appropriate response may not be made when real problems with drug safety do occur.

Finally, despite these considerations, it should be pointed out that, in the end, the system did work in New Zealand, the Minister and the Department of Health generally handled the controversy well, and the right decisions were eventually made. What is important now is that the system of assessment of adverse drug reactions is itself assessed to ensure an optimal response when such problems occur again in the future, as they inevitably will.

Acknowledgements

Neil Pearce is funded by a Senior Research Fellowship of the Health Research Council of New Zealand.

References

Bernier, R. (1990) Asthma drug controversy climaxes with government decision to restrict use *Epidemiology Monitor* 11 (3): 1-5.

Buist, A.S., Burney, P.G.J., Feinstein, A.R. et al. (1989) Fenoterol and fatal asthma *Lancet* i: 1071 (letter).

Buist, S., Burney, P.G.J., Ernst, P. et al. (1989) *Consensus report: an appraisal of a manuscript by N. Pearce et al.* Los Angeles, Boehringer Ingelheim.

Bown, W. (27 July 1991) Warning letter links asthma deaths to drugs *New Scientist* p. 9.

Crane, J., Pearce, N., Flatt, A. et al. (1989) Prescribed fenoterol and death from asthma in New Zealand, 1981-1983: a case-control study *Lancet* i: 917-22.

Editorial (1972) Asthma deaths. A question answered *British Medical Journal* ii: 443-4.

Elwood, J.M. (1989) *Prescribed fenoterol and deaths from asthma in New Zealand — Second report* Wellington, Department of Health.

Elwood, J.M. and Skegg, D.C.G. (1989) *Review of studies relating prescribed fenoterol and death from asthma in New Zealand* Wellington, Department of Health.

Esdaile, J.M., Feinstein, A.R. and Horwitz, R.I. (1987) A reappraisal of the United Kingdom epidemic of fatal asthma *Archives of Internal Medicine* 147: 543-9.

Grainger, J., Woodman, K., Pearce, N.E. et al. (1991) Prescribed fenoterol and death from asthma in New Zealand 1981-7: a further case-control study *Thorax* 46: 105-11.

Hendeles, L. and Weinberger, M. (1983) Nonprescription sale of inhaled metaproterenol — déjà vu *New England Journal of Medicine* 310: 207-8.

Hodge, J.V. (1989) *Fenoterol and asthma mortality: report on attendance at a meeting in Los Angeles on 29 July 1989* Auckland, Medical Research Council of New Zealand.

Jackson, R.T., Beaglehole, R., Rea, H.H. et al. (1982) Mortality from asthma: a new epidemic in New Zealand *British Medical Journal* 285: 771-4.

Lanes, S.F. and Walker, A.M. (1987) Do pressurized bronchodilator aerosols cause death among asthmatics? *American Journal of Epidemiology* 125: 755-60.

Leeder, S.J. (21 December 1988) *Minutes of a meeting of asthma investigators to discuss progress of a study of asthma deaths in New Zealand and their association with the use of inhaled fenoterol* Wellington.

Olson, L.G. (1988) Acute severe asthma: what to do until the ambulance arrives *New Ethicals* 105-16.

Pearce, N.E., Crane, J., Burgess, C., and Beasley, R. (1990a) Study designs for examining death from asthma: the case-control approach. In Ruffin, R.E. (ed) *Asthma mortality: proceedings of the second national asthma mortality workshop* 23-6 Sydney, Excerpta Medica.

Pearce, N. E., Grainger, J., Atkinson, M. et al. (1990b) Case-control study of prescribed fenoterol and death from asthma in New Zealand, 1977-1981. *Thorax* 45: 170-5.

Pearce, N.E., Crane, J., Burgess, C. et al. (1991) Beta agonists and asthma mortality: déjà vu *Clinical and Experimental Allergy* 21: 401-10.

Rea, H.H., Scragg, R., Jackson, R. et al. (1986) A case-control study of deaths from asthma *Thorax* 41: 833-9.

Sackett, D.L., Shannon, H.S., and Browman, G.W. (1990) Fenoterol and fatal asthma *Lancet* i: 46 (letter).

Savitz, D.A., Greenland, S., Stolley, P.D. et al. (1990) Scientific standards of criticism: a reaction to 'Scientific standards in epidemiologic studies of the menace of daily life' by A. R. Feinstein *Epidemiology* 1: 78-83.

Sears, M.R., Rea, H.H., Beaglehole, R. et al. (1985) Asthma mortality in New Zealand: a two year national study *New Zealand Medical Journal* 98: 271-5.

Sears, M.R. and Rea, H.H. (1987) Patients at risk for dying of asthma: New Zealand experience *Journal of Allergy and Clinical Immunology* 80: 477-81.

Sears, M.R., Taylor, D.R., Print, C.G. et al. (1990) Regular inhaled beta agonist treatment in bronchial asthma *Lancet* 336: 1391-6.

Speizer, F.E. and Doll, R. (1968) A century of asthma deaths in young people *British Medical Journal* 3: 245-6.

Speizer, F.E., Doll, R., and Heaf, P. (1968) Observations on recent increase in mortality from asthma *British Medical Journal* i: 335-9.

Spitzer, W.O., Suissa, S., Ernst, P. et al. (in press) Beta agonists and the risk of asthma death and near fatal asthma. Presented at the International Symposium on Pharmacoepidemiology, Basel, August 1991 (abstract).

Stolley, P.D. (1972) Why the United States was spared an epidemic of deaths due to asthma *American Review of Respiratory Disease* 105: 883-90.

Stolley, P.D. (1989) A public health perspective from academia. In Strom, B.L. (ed) *Pharmacoepidemiology* New York, Churchill Livingstone.

Venning, G.R. (1983) Identification of adverse reactions to new drugs. I. What have been the important adverse reactions since thalidomide? *British Medical Journal* 286: 199-202.

Wilson, J.D., Sutherland, D.C., and Thomas, A.C. (1981) Has the change to beta agonists combined with oral theophylline increased cases of fatal asthma? *Lancet* i: 1235-7.

Withdrawal of the Copper 7: The Regulatory Framework and the Politics of Population Control

Phillida Bunkle

The market for pharmaceuticals designed for well populations is potentially greater than that for sick ones. The market for contraceptives is therefore large, involving as it does most adult women for much of their adult lives. In New Zealand this market is ambivalently placed between the third and first world. The health system is close to that of developed capitalist societies, and provides an infrastructure that facilitates the consumption of pharmaceutical products, yet the regulatory structure is under-developed and consumer protection is legally weak. This suggests that the population is segmented partially along race, class, and gender lines into pockets of first and third world consumers, which makes New Zealand women vulnerable to product testing, particularly as the Accident Compensation scheme provides a ready insurance against liability for manufacturers of new products.

It is argued in this chapter that contraceptive manufacturers and their representatives protect their markets, by supporting the United States population control lobby, by pressing for funding of research and supply programmes and for the weakening of the regulatory system, and lobbying for the repeal of prohibitions on the international distribution of products not approved for sale in the United States. This chapter will show that the choices available in New Zealand have been vitally affected by these activities. It is argued that the direction of contraceptive research and development, and hence the methods which become available throughout the world, are dominated by the priorities of the American population control establishment.

The major legal cases that have developed around contraceptive safety in the United States provide a unique opportunity to see inside major transnational pharmaceutical firms. The discovery proceedings provide extensive documentation of the activities of these companies, enabling observers to describe in detail how medical knowledge is manufactured. This chapter will examine documents made public through cases against G.D. Searle Ltd of Illinois, the makers of the Copper 7 intra-uterine device (IUD), in order to describe some of these processes, particularly the integration of the State with manufacturing and population control interests.

The Early History of IUDs

Until the mid-1960s the major funders for contraceptive research and development were the pharmaceutical companies. In 1965 the companies funded 32 per cent of world-wide research. By 1979, 58 per cent was spent by the United States Government and only about 9 per cent was spent by the companies. This remarkable change was achieved by the population control lobby which persuaded the US Government that population growth was a matter of strategic interest. By 1982 the United States was 'the single largest provider of international population assistance' and its contribution amounted to 'almost one-half of all international population assistance' (Johns Hopkins University, 1983).

The Population Council, initiated in 1952 by John D. Rockefeller III, 'seized on the IUD as the one device with the potential for solving the impending population crisis in the developing nations' and spearheaded the development of IUDs, sponsoring conferences and research 'and even funding several of the new designs. The rights to manufacture the new devices were quickly bought up by pharmaceutical companies eager to enter this new market' (Pappert, 1986).

The Council organized and funded the First International Conference on Intra-uterine Contraception in New York in 1962. Enthusiasm for developing the IUD as the technological solution to controlling third world populations was high.

Jack Lippes, progenitor of the Lippes loop IUD, was a consultant to both the Council and to Ortho Pharmaceuticals, a subsidiary of Johnson and Johnson Ltd, who started marketing the loop in 1962. The loop remained the most popular brand through the 1960s. The Dalkon Shield went on the market in 1971. It was often inserted with a local anaesthetic administered to the cervix. The Shield was marketed in two sizes and promoted directly to American nulliparous women (that is, women who had not borne children). This promoted fierce competition for the home market.

In the late 1960s the Population Council employed Dr Howard Tatum to develop the Copper T IUD, while another worker, Zipper, was developing the Copper 7. Both hoped their designs — shaped like the letter T and shaped like the figure 7 respectively — could be easily fitted without an anaesthetic in nullipara, would be more effective in preventing pregnancy, and less likely to be removed. A wave of unrestrained inventiveness followed as loops, bows, coils, rings, spirals, and springs were trialled on unsuspecting women, and the Population Council set up a Cooperative Statistical Program to study them (Tatum, 1972).

Initially doctors feared that IUDs would introduce infection into the uterus from the vagina. Once they were widely used, medical studies demonstrated that this was in fact the case. Despite warnings in the medical literature, development of the IUD went ahead very rapidly. The studies, which indicated that IUD users were at risk from infection, were frequently swamped by data generated by the companies which had developed the devices and distributed them. An article by Professor Hugh Davis on the safety of the Dalkon Shield was distributed by manufacturers A.H. Robbins long after its claims were known to be invalid (Mintz, 1985). A massive promotion campaign was launched by manufacturers in the late 1960s and early 1970s to persuade doctors that the IUD was safe. Doctors were taught to reassure women that severe cramps, bleeding or discharge were normal after insertion (Kasindorf, 1980: 28).

In New Zealand many doctors apparently believed that pain on intercourse, which might be a symptom of infection, was a neurotic symptom which should be treated with reassurance or tranquillizers. Numerous women have reported to consumer advocacy groups that when they described such symptoms they were not given any physical examination at all.

Searle Develop the Copper 7

Adverse information about the oral contraceptive pill — the Pill — was highly publicized in the late 1960s, and profits from that product began to fall. In 1965 oral contraceptives accounted for 44 per cent of G.D. Searle Ltd's sales. In 1968 this began to fall; by 1970 it had dropped to 17 per cent and by the mid-1970s oral contraceptives amounted to only 6 to 8 per cent of total sales revenue (Searle, 1968–1976).

Zipper's IUD design, shaped like a figure seven, was developed by Searle, and its efficacy was enhanced by copper wire wound around its stem. The company claimed that the copper leached into the tissues of the womb, but no one knew how or why it had a contraceptive effect (Zipper, 1969). The Copper 7 could be folded into an insertion tube sufficiently narrow to pass through a woman's cervix without prior dilation or stretching of the cervix. The company maintained that it could even be inserted into nulliparous women.

On 6 November 1970 a Searle official wrote: 'Our initial aim will be to have as many devices as possible inserted quickly so that we can start accumulating data on as many patients as possible'. The clinical trial needed many investigators 'in order to accumulate the patients at a faster rate', but, he reassured doctors, 'we fully realize that this will be your first direct experience with an IUD, but you are not alone, it's ours too' (Searle, 1970).

Even before the safety trial began, the Medical Department at Searle had already concluded 'that efficacy and safety data to date are adequate for medical release for marketing'. The Minutes of 25 November 1970 of Searle's Product Introduction Committee continued, 'We will run two short studies under our own monitorship, as fast as they can be started, and data are to be called in so that a final medical conclusion can be reached by April 1st, 1971'.

Medicines usually face rather more than six months of testing, but the Food and Drug Administration (FDA) regarded IUDs as devices, not drugs. In December 1970 Searle began recruiting doctors to be Clinical Research Associates in a world-wide study. On 22 March 1971, however, the FDA responded to the claim that the copper was an active ingredient and re-classified the Copper 7 as a prescription drug. The product then had to face pre-clinical animal tests as well as longer, more rigorous clinical testing on humans. This allowed the all-plastic Dalkon Shield, which did not require pre-marketing approval, to reach the market first. It went on sale a few months later.

Eventually Searle was to claim the FDA's approval as a major selling point, but in 1970, when faced with these delays, the company decided not to wait for FDA approval but to turn to less-regulated overseas markets. On 10 May 1971 a Searle memo from Dr Gibor on international marketing of the Copper 7 said that it had been:

... decided to initiate a crash program for studying the Cu7 to collect sufficient data, as soon as possible, to be able to market the Cu7 internationally before the Population Council markets the Cu T. We feel confident now that within the next 2 to 3 months we will have sufficient data to support international marketing of the Cu7.

They believed that they 'could supply a very good story to back the Cu7' and reached the conclusion that 'we could get into the international market within 4 to 5 months'.

Some of the advantages listed for such precipitate introduction of the Copper 7 to the international market were 'to beat the possible competition of the Copper-T in a market unprotected with our U.S. patent', that 'a wide acceptance of the Cu7 internationally may influence the FDA to look at it more favourably' and 'last, but not least, it is time to see some profits on our investments in this project'.

Searle was aware that the rush could compromise the clinical trials. Dr Gibor wrote in a memo: '. . . the ultimate objective is to file the NDA [New Drug Application] by August 1, 1971, even if this will mean sacrificing some of the medical and scientific aspects of our studies' (Searle, 20 April 1971).

The Clinical Trials

Lawyers for the women in the Copper 7 case have alleged that the clinical trials had serious deficiencies. The doctors recruited as Clinical Research Associates were given single page forms to fill out for each woman entering the trial. Follow-up forms were periodically filled out until the device was removed, or the woman could no longer be traced, or the study ended (Investigator's Manual for Copper 7 Device for Intrauterine Contraception).

It is alleged that 15.5 per cent of the Canadian subjects were lost to follow-up as little as two months after insertion, and in the US 15.6 per cent were lost after a year. The trials claimed to show a very low rate of Pelvic Inflammatory Disease (PID), but it seems that those lost to follow-up were not deducted from the total when the rate was calculated.

Women who had the device removed ceased to be subjects in the studies, and nothing is known about their subsequent health; yet it was the women who had it removed for medical reasons whose problems were most likely to be serious. Later in litigation it was disclosed that Copper 7 devices were removed from one million American women in the first year after insertion and that the removal rate after two years was 19.1 per cent and after three years 25.9 per cent (Mintz, 1988).

Infertility as a side-effect was inadequately investigated in the Report of the Clinical Trials. The Report merely said that: 'Data show that, of the 64 women who are known to have become pregnant after removal of Cu7, 57 (89%) conceived within 6 months of device removal' (G.D. Searle, 1972). This ignored the fact that 30 per cent of those who had the device removed to conceive did not become pregnant. Searle promised that 'patients who drop from the study to become pregnant will be followed to document the return of fertility after removal of the Cu7. Patients who do not become pregnant will be followed in an attempt to determine the cause of their infertility' (Mintz, 1989). However, the company later conceded in evidence in legal proceedings that this was never done.

The incidence of PID was the crucial safety issue, yet the follow-up form filled out by investigating doctors did not have PID as an option which could be recorded. Nor could the doctor record the combination of symptoms that might characterize pelvic infections, because the form said 'check only one response' (Investigator's Manual for Copper 7 Device for Intrauterine Contraception, form ICU 7F).

Two New Zealand doctors were recruited as Clinical Research Associates in the clinical trials. Dr Baeyertz of Wanganui enrolled 23 women — none expelled the device or became pregnant, and one had it removed for medical reasons. Dr Jarvis of National Women's Hospital enrolled 18 women — one expelled the device but none became pregnant or had a medical removal (New Drug Application, 1972).

This safety data is not available for public appraisal. Whether it was ever submitted to the New Zealand Health Department is unclear. Drug firms own the results of safety tests on their products and do not have to publish them or make them available for independent assessment; nor do participating researchers in public hospitals.

No Departmental file records when the Copper 7, known here as the Gravigard, appeared on the New Zealand market, but it is believed that it was first promoted in New Zealand in late 1974 and 1975. Dr Bob Boyd, head of the section of the New Zealand Department of Health which regulates therapeutics, does not know if a definite decision was ever made to classify copper IUDs as medicines, although he says it 'seems as if' the Department began treating the Gravigard 'as if it was' in 1978.

By contrast, in Australia, the Australian Drug Evaluation Committee (ADEC) classified the copper IUD as a 'therapeutic substance' and the Australian Department of Health carried out an extensive review of safety data. This included chemistry, quality control data, and human clinical and animal pharmacology data, including toxicology and carcinogenicity studies. In addition, the information on package inserts, prescribing information, and promotional material were examined (ADEC, 1974).

The Tailstring

There was a history of unreported PID associated with IUDs. In May 1967 three deaths as a result of PID were reported in the United States (Kasindorf, 1980). All three women had worn IUDs. The FDA began to take an interest but did not actively regulate the devices. The next year a survey of serious complications was published in *Obstetrics and Gynaecology* (Scott, 1968). Max Elstein, a British researcher, questioned the use of IUDs in nulliparous women (Elstein, 1967). In August 1968 Dr Nicholas Wright published a study which showed that a woman wearing an IUD had a six times greater risk of getting PID than a woman who did not (Wright and Laemmle, 1968).

In 1972 Dr Michael Burnhill presented a paper to Planned Parenthood which was published later that year. He explained that 45 per cent of women wearing IUDs suffered from chronic inflammation of the lining of the uterus and that this was often accompanied by an unpleasant discharge. His study suggested that infection could enter the uterus via the tailstring months after insertion. The population lobby, however, were unwilling to deal with his findings. He commented:

It was put down as being anecdotal, nonscientific and not relevant . . . I was describing the syndrome carefully, and the description is as true today as it was then. The family planners were not ready to deal with the relationship between infection and the IUD (Kasindorf, 1980).

Nevertheless, it was the mounting evidence of damage, especially of infection and its consequences, that led to the withdrawal in 1974 of a widely marketed IUD, the Dalkon Shield, which had been released only three years before. The most publicized design defect was alleged to be the multifilament tailstring which allowed the 'wicking' of fluid containing bacterium from the vagina into the normally sterile uterus (Mintz, 1985).

After the Dalkon Shield was removed from the market, Searle made a major advertising push for the Copper 7 in US medical journals. Sales were astonishing and within a year it had an 80 per cent market share. There were, however, already grave concerns being expressed within Searle about the quality and safety of the Copper 7's composition and design. The tailstring was advertised as a monofilament but its polypropylene composition was nevertheless subject to fraying and serious surface deterioration.

A few months after the FDA had finally approved the Copper 7 for marketing in the US in February 1974, Dr John Vance, Medical Director of Searle in Canada, wrote a memo to Searle USA about his impression of magnified pictures of the tailstring. He wrote, it 'looks like a dog's breakfast after the rats have been at it . . . the surface is rough and irregular and the ends of the copper wire look like a mediaeval weapon' (Vance, 1974).

In October 1975, when the Copper 7 was being heavily promoted in American medical journals, a company representative in Texas sent back a Copper 7 on which the string had broken upon removal, and reported a conversation with another doctor who was very concerned about the incidence of PID (Brown, 1975). Searle's Dr O'Brien replied that 'the string breakage problem came to our attention last year when investigational devices were being removed' and claimed that a 'stronger line is now being used' (O'Brien, 1975).

In 1975 another Clinical Research Associate doctor gave a conference paper documenting another sort of problem. The tailstring was bent over the top of the insertion tube and ran down the outside. Polypropylene had a 'memory' and after insertion would sometimes return to this bent shape pulling the string up into the uterus, which would increase the likelihood of infection and make it impossible to check the presence and position of the device.

There was no apparent internal response to this problem until 1977, when O'Brien said to a manager for regulatory affairs at Searle, 'We have been aware of this problem for the last 18 months and have been seeking a satisfactory solution' (G.D. Searle, 1977). A manager wrote on the memo, 'We have a design problem . . .'

In March 1979 a Searle technical manager at the manufacturing site in England reported that examination with the naked eye revealed 'fronds' on many tailstrings, even new ones. 'We need advice urgently on the amount of fronding to be permitted . . . I assume that Medical [Searle's Medical Department] are not worried about these fronds . . . We have obviously had them for years on some strings' (Morepeth Plant, 1979). This time the company acted swiftly, quality control investigated the matter and concluded that fraying is 'inherent in the filament micro structure' of polypropylene. A week later they recommended a change to polyethylene, since 'fraying of some degree exists throughout our production'. The company made plans to change the string. 'A string free from user/physician complaints (visibility, penile irritation, fraying, memory) should be available for marketing by December 1981' (G.D. Searle, 1980).

By this time there was concern in Australia about the quality of the tailstring. The Commonwealth Department of Health commissioned the National Biological Standards Laboratory to investigate complaints of fraying. In its 1980 report the Laboratory noted that it seemed that the tailstring was made of 'poor quality as normal commercial quality . . . filament is smooth on the surface and free from the above defects' (Report of the National Biological Standards Laboratory, February 1980).

There is no evidence of any concern by the New Zealand Department of Health until two years later when Searle applied to change the threading system, so that 'we can run the string down the inside of the tube and eliminate this looping problem. It should also provide for a more sterile insertion procedure' (Hilsden, 1982). Doctor Boyd wrote on the Drug Action Sheet — 'Good, about time. It was suggested to them ages ago'. However, the Department did not actively follow up this issue, and took no further interest until a further four years had passed.

The Size of the Copper 7

In 1967 Max Elstein had published an article questioning the use of IUDs in women who had not had children (Elstein, 1967). The common fear was that young or nulliparous women with small uteri would have difficulty using the IUD, with a high incidence of difficult insertions, pain, and expulsion, and that these women were especially vulnerable to infection and to infertility as a result of infection.

Even before clinical trials began on the Copper 7, corporate executives were aware that their device might not be suitable for nulliparous women and were talking about the need to develop a 'distinct separate product' for them (G.D. Searle, 9 November 1970). Although the insertion tube was narrow, the vertical arm was 3.6 cm, which was 0.8 cm longer than the uterine cavity of the average white American woman who has not had a baby. In 1973 one of the Clinical Research Associate doctors expressed the view that 'changing the dimensions of the Cu7 would reduce significantly the expulsion, pregnancy and removal rates' (G.D. Searle, 1 July 1973).

On 24 June 1974, as the marketing of the large device was about to begin, Searle's director of clinical obstetrical and gynaecological research wrote in a memo: 'Since we do not know whether the present Cu7 model is the correct size and since it appears likely that it is not the ideal size for the small nulligravid uterus we must investigate the concept of a smaller device and possibly market two sizes'.

The next year, in 1975, Searle distributed a 'Dear Doctor' letter to New Zealand doctors describing the Copper 7 as 'extremely small so that it can be used in nulliparous women. Also there is minimal distension of the uterine cavity with extremely [low] incidence of bleeding and discomfort'. But by July of that year Searle had in fact developed a proposal for a smaller device. This proposal is one of the most incriminating documents discovered during litigation because it shows that the company was aware of evidence showing that the people to whom it was selling its product were those most at risk of injury from it. The reason for developing the small device was that the majority of their customers were the young, white, nulliparous women for whom it was too large. 'The age of the large majority of Cu7 users is between 20 and 25 (72%). The great majority are white and about 40% have never been pregnant' (Line Extension Proposal, 1975).

At this time the Copper 7 had 80 per cent of the US market, and the company predicted that a small device would expand the market even further: 'The small size could totally dominate the market for new insertions within a year or two. More importantly it would prevent loss of market share to a competitor with a small size' (Line Extension Proposal, 1975). The Proposal also revealed how profitable the IUD was. Profits were 80 per cent of gross sales, even including the costs of testing and promotion. For the small device the 'capital expenditures are expected to range between $32,000 and $38,000 . . . In the first year of marketing the smaller size seven is expected to bring in $1,400,000 gross . . . Cumulative sales figures for the small seven in five years should be nearly $20,000,000' (Line Extension Proposal, 1975). The company developed a new small device.

The Mini-Gravigard in New Zealand

The company wrote to the New Zealand Department of Health 'advising our intention to market . . . The Mini-Gravigard is designed for use by nulliparous women and by women with uteri thought to be too small to accommodate the standard frame device' (James, 1979). Once again the safety studies submitted to the Department are not available to the public. The Health Department treated it as a change to an existing drug, but asked for more up-to-date or longer clinical trial results, and inquired about marketing status in other countries (Giffith, n.d.).

The company replied that 'Marketing submissions have not yet been lodged in the U.S., U.K., Australia or Canada.' They said that longer clinical trials were underway in Europe and Australia and that the results would be submitted when they were available (Coy, n.d.). It is not possible to tell from the censored record if this was ever done. The Department did not follow up on these unsatisfactory answers to their questions and two weeks later gave permission for the Mini-Gravigard to be marketed in New Zealand almost immediately. This was followed in June 1980 by an application for the Mini-Gravigard to be included in the Drug Tariff (Villers, 1980). Once again the procedure seems to have been perfunctory.

The basic information was clinical trial data, which cannot be accessed or evaluated by the public. The submission by the company did not mention any trials carried out in New Zealand, but listed Dr Jennifer Wilson, Dr John Taylor, and Dr T. Fiddes as local practitioners who had experience with the device (Pharmaceutical Benefits Mini-Gravigard Submission by G.D. Searle (NZ) Ltd). There is no evidence that the Department consulted them, but Jennifer Wilson volunteered her experiences with 27 insertions. Her data was limited but she concluded: 'On these small numbers I would say that the performance of the device is not good. I could not support the small frame copper being registered as there is a perfectly good alternative . . .' (Wilson, 1980). The Department acknowledged her letter on 19 August 1980, and sent it to the Pharmacology and Therapeutics Advisory Committee, which accepted the Mini-Gravigard on the Drug Tariff the next day (Combly, 1980).

Once again the Australian regulatory process was more cautious, even if it was no more effective in protecting the consumer. Searle applied to the FDA for marketing approval for the Mini-Gravigard in February 1981. In June of that year they also applied to the Australian Department of Health. In August ADEC resolved that there should be no objection so long as chemistry, quality control, and clinical trial data were satisfactory (ADEC, 1981). The Department strongly encouraged the company to provide more forthright prescriber information. ADEC granted approval in December 1981.

A year later, by which time Searle estimated that it had inserted half a million Gravigard and Mini-Gravigard into Australian women, the Department had second thoughts. ADEC noted that approval had been granted without full evaluation of the safety and efficacy of data on the Mini-Gravigard, and in December 1982 a decision was made to undertake such an evaluation (ADEC, 1982). The results, however, did not come before ADEC for nearly three years. In 1985, when they were forthcoming, the evaluation was very critical of the safety data derived from the clinical trials, maintaining that it did not substantiate the safety and efficacy of the Mini-Gravigard in nulliparous women:

Although the data presented are largely inconclusive, it is possible that the high incidence of adverse reactions of discomfort and bleeding in 50 to 70% of participants, short average experience of wear of the device per woman, and a possible high pregnancy rate, corroborate opinion that intra-uterine devices are not first choice contraception for nulliparous women.

They concluded that there was therefore no identifiable group who would specifically benefit from the Mini-Gravigard, as the company claimed, and even more significantly, 'it appears that data presented do not in fact resolve questions on the safety and efficacy of Mini-Gravigard' (ADEC Resolution No. 2300).

In February 1986, one month after the Copper 7 was removed from the US market, the Australian Department of Health stated that until the clinical trials of the Mini-Gravigard were concluded and evaluated, the Mini-Gravigard 'should have clinical trial status until such time as adequate data have shown safety and efficacy' (Australian Department of Health, 1986). Unfortunately there is no evidence that the Department acted on their own advice, and the Mini-Gravigard continued to be sold as freely in Australia as in New Zealand.

In February 1981, 16 months after giving notice of intention to market the Mini-Gravigard in New Zealand, Searle made a full marketing application to the FDA for approval to market the Mini-Gravigard in the United States, where it encountered more serious regulatory scrutiny.

Withdrawal from the Market

In September 1982 a Searle executive reported to the company a meeting with FDA officials who informed him that the FDA itself intended to pose a series of searching questions to the Fertility and Maternal Drugs Advisory Committee when it met on 28 October to consider the New Drug Application for the Mini-Gravigard (Searle, 14 September 1982). The Committee included a number of practising specialists whose experience made them especially wary of directing a product at the vulnerable group of nulliparous women. The FDA officials noted that the drop-out rate of the clinical trials was 'exceedingly high . . . only 59% of U.S. patients still had the small Cu7 in situ at the end of the study'. Their conclusion was that 'the data presented in this application were not derived from well designed, well controlled, adequate studies upon which one could conclude that there was substantial evidence that the small Cu7 was both safe and effective'. When the Committee took an adjournment it was clearly not convinced of the safety of the device (Fertility and Maternal Health Drugs Advisory Committee, 1982). At this point Searle withdrew its application. It is possible that an outright rejection by the FDA might have hurt their well-established overseas markets in countries such as New Zealand, Australia, and the UK, where sales continued uninterrupted.

The Mini-Gravigard was never sold in the US, where Searle continued its 'one-size-fits-all' sales strategy, continuing to claim that the large device was suitable for nullipara. New Zealand women had a choice of size, but again proved vulnerable to the marketing of devices that have not been accepted in the US.

Searle did not, of course, draw attention to the fact that the FDA had not approved the Mini-Gravigard. Indeed, sometimes they appear to have perpetuated an assumption that both were approved. In January 1986 Searle withdrew the Copper 7 from the US market.

It took three years of lobbying by consumer advocacy groups to determine that sales cease in New Zealand. An IUD Advisory Committee was established in 1986 by the Department of Health, largely as a result of consumer advocacy groups pressing for decisive action to institute proper regulatory control of devices and to inform women clearly of the risks. The Committee reviewed IUDs and noticed that the string remained on the outside of the inserter tube four years after Searle had applied to change it. On 6 June 1986 the Department, prompted by the Advisory Committee, asked Searle (NZ) why these changes had not been made (Nisbet, 6 June 1986). Three months later they asked again (Nisbet, 17 September 1986). In October, Searle responded by challenging the necessity of the changes and suggesting that the removal of the Copper 7 from the US market was not because of valid concerns over safety but rather due to 'unwarranted product litigation' in the US (Van de Water, 20 October 1986).

In October Searle ceased to manufacture the Copper 7 worldwide, but continued to distribute it in New Zealand. The Department responded by writing to the company asking them to 'consider' discontinuing sales (Nisbet, 27 November 1986). The Company replied that it had given consideration to the suggestion but that: 'It is our feeling that the most appropriate way to handle this unfortunate situation is to continue marketing both devices until stocks are exhausted' (Van de Water, 10 December 1986). The Department annotated the letter 'accept this proposal . . . but please recheck in March '89 to ensure stocks are exhausted'. The Department appeared wary of using its power to demand that sales stop. The only instrument that the Health Department felt confident to use was its control of the Drug Tariff; it recommended that both IUDs be removed from the Drug Tariff from 1 February 1987 so that they were no longer free prescription items (Riseley, 19 December 1986).

In 1987 Searle made plans to recommence manufacture of both devices and was keen to reinstate them (Van de Water, 9 November 1971). The Department, prompted by the IUD Advisory Committee, again raised the question of the composition and position of the thread, and questioned the suitability of the device for nulliparous women (Nisbet, 11 December 1987).

Six months later the managing director of G.D. Searle (NZ) Ltd answered: 'I had hoped to have received a response to your enquiry from our International Medical Affairs Department, but my request seems to have fallen between the cracks' (Van de Water, 6 May 1988). In June

1988 he provided 'information and comments prepared by our head office in Chicago' (Van de Water, 23 June 1988). Unfortunately, the Department of Health will not release these Searle documents for evaluation. The data sheet, however, continued to describe the tail as polypropylene.

On 4 October 1988 the Department told Searle that the IUD Advisory Committee was not satisfied and still wanted the composition and location of the string changed (Nisbet, 4 October 1988). Nine months later in July 1989 Searle replied, saying that they would address these issues: 'We believe that there are sufficient data to support continued use of the current design. In addition Searle believes that, if Directions For Use are properly followed by the physician, string looping and partner discomfort should not occur' (Van de Water, 19 July 1989). Once again the information they supplied has been withheld from the public. Searle also supplied a list of countries where the devices were 'approved for marketing', which included New Zealand. The Advisory Committee responded that 'It was of surprise to find that the list stated that it was marketed in New Zealand' (Nisbet, 14 August 1989). Searle pointed out that while it had been withdrawn from the Drug Tariff, approval for sale had never been withdrawn (Van de Water, 18 August 1989). In August 1989 Searle accepted that neither IUD would be accepted back on the Drug Tariff without design modifications, which they were not prepared to make, and in September 1989 Searle finally agreed not to distribute any more (Van de Water, 21 September 1989).

Issues and Consequences

Approval to sell the device in New Zealand had still not been officially withdrawn. Had it been, recall, that is the removal of devices from women still wearing them, would have been an issue.

Twelve years after the first internal memo accepted the need for change, and nine years after it was acknowledged to cause serious medical problems, the company was still defending its unchanged composition and mode of insertion. Some clue as to why may be gleaned from two internal Searle memos produced during litigation. One said, 'Hold-up on IUDs in legal cases — Legal Department does not want us coming out with "New and Improved" product design . . . this is slowing everything down regarding IUDs' (Loxley, n.d.). The second says, 'It does not look like we will be "allowed" to pursue string changes'. This memo throws some light on why such an important corporate decision has left apparently few traces when it adds 'keeping all this *out* of writing is a difficult task' (G.D. Searle, 14 May 1985).

Discovery proceedings have identified a communication between two of Searle's experts on toxicology about an arrangement organized by Searle's general counsel, for litigation purposes 'that would offer even greater insulation of both you and me . . . this would be for theatrical purposes and would not in any way change the way in which we are currently operating'; it would, however, 'offer us maximum insulation'. The writer added, 'In addition, this memo should self-destruct' (Searle, April 1975).

There is only one letter on the Health Department file from a New Zealand consumer who had a Copper 7 inserted by a specialist in 1973: 'Within 24 hours I had become ill with a high fever, this was the beginning of 27 months of great discomfort and distress'. She was treated for infection 13 times before the doctor finally removed the device. 'I feel confident', she said, 'that there must be many women who suffered similar problems' (Anonymous, 29 October 1985). By September 1991 nearly 200 women who believed the Copper 7 had injured them had come forward. No one knows how much these injuries have cost.

Westrom estimates that in Australia PID causes '5600 new cases of infertility each year to be added to the already existing pool of infertile women'. There are 33,000 cases of PID in Australia a year which has contributed to the sevenfold risk-increase of an ectopic pregnancy. 'One woman out of four will suffer from one or more of chronic abdominal pain, infertility, or ectopic pregnancy after one or more episodes of PID' (Westrom, 1980; Westrom and Maardh, 1984). IUD use exacerbates the PID problem (Senanayke and Kramer, 1980; Edelman, Berger, and Keith, 1982).

In the United States it is estimated that 250,000 women have become sterile from using IUDs. Some 200,000 women a year develop PID, many related to the use of an IUD (McDonnell, 1986).

Towards a Feminist Analysis

Contraception was slow to be made widely available in New Zealand, and the birth rate remained high compared to European and North American societies (Vosburgh, 1978). In the 1930s a consumer group, the Sex Hygiene and Birth Regulation Society, was established to challenge the medical monopoly of information and to take direct action on access to contraceptive services. In 1939 this became the Family Planning Association (Fenwick, 1980). By 1961 the Family Planning Association had compromised with the profession and access to contraception was eased for married women, large numbers of whom began to use the oral contraceptive pill when it was introduced in the early 1960s (Fenwick, 1980; Else, 1991). The availability of contraception for some groups of women heightened the tension for others, such as the unmarried, from whom it was withheld and who were in the 1960s increasingly subject to contradictory sexual expectations. In the early 1970s more militant groups began to demand access to fertility control (Bunkle, 1988). Their arguments were couched in terms of each woman's right to choice and to control of her own body, rather than the earlier justification of family welfare or the stability of social institutions (Dann, 1985).

The analysis developed by the militant groups in the 1970s showed that in a consultation about contraception, both the practitioner and patient were subject to competing and contradictory pressures determined in part by class, race, and sexual mores. For both parties, sexual function and fertility involved core issues of personal identity. Practitioners found themselves caught between complex claims. The militant women's health movements of the 1970s successfully challenged the right of the profession to use the power of the prescription pad to enforce their own expectations about appropriate sexual behaviour. This movement to change attitudes has been largely successful. Professional attitudes are no longer a significant barrier to access to contraception, although they may determine the methods used. By the 1980s contraception had become readily available and widely used.

Arguing for access to contraception implied endorsement of the methods available, which were generally accepted as unproblematic by professional and consumer groups alike. But wide-scale use in the 1960s and 1970s eventually revealed significant hazards. However, apart from a British report citing the risk of serious circulatory problems associated with the Pill (Royal College of General Practitioners, 1974), information about side-effects tended to remain anecdotal and largely invisible to practitioners. The women's health movement responded more quickly, and concerns about safety became more prominent than concerns about access. In the 1980s the women's health movement moved from a focus on choice to an analysis of how those choices were constructed, and how those constructions limited the options available to individual patients (Bunkle, 1984).

This analysis led to a far more wide-reaching analysis of the pharmaceutical industry, and of its influence on medical research, on the regulatory process, and on the production of a climate of opinion receptive to its wares. It entailed examining the market dynamic of the pharmaceutical industry and its relation to governments and regulatory processes, both globally and locally.

In analysing the construction of medical information about contraceptive drugs and devices, feminists have pointed to the high level of integration of the profession, the State, population control organizations, and the industry, and the involvement of the latter in producing ostensibly objective and scientific information, in designing and administering regulations, and in making clients more receptive to their products. This integration prevents the medical profession and the State from acting independently to protect the interests of consumers.

The women's health movement questions the medical profession's belief in the objectivity and scientific validity of its knowledge base, and challenges its unexamined reliance on sources of medical information and processes of regulation which can be contaminated by commercial interests. It is hardly surprising that the construction of medical information is not obvious to doctors, because so much of it takes place within the secrecy of the corporation. This secrecy is protected by the regulatory system. It is especially important that this process be examined now when deregulation is fostering closer ties between the market and the production of information.

Conclusion

The Copper 7 case raises again the adequacy of regulatory procedures in New Zealand. The regulatory process consists of a dialogue between the Department and the manufacturer without any mechanism for input from consumer interests. The Department is dependent on the information given to it by manufacturers, who have a vested interest in the outcome. The Department is clearly too under-resourced to take on an active, investigational role. There is no evidence on the file that it undertook an independent evaluation of the Copper 7 Clinical Trial Reports.

Many doctors uncritically accepted the scientific validity of the clinical trials. Neither doctors nor independent scientists have access to the original data, which is created and owned by the manufacturers. Advertising claims sometimes appear to have been accepted at face value, and their basis is not accessible for evaluation. Scientific objectivity rests on the possibility of refutation and therefore on open debate. But the demands of commercial secrecy place the crucial data beyond scrutiny.

While individual doctors were quick to observe, and some to report, adverse findings, the company dispensed with their letters. The company had no motivation to investigate practitioners' concerns, and there was no procedure in the regulatory system by which they could be followed up. In New Zealand the Medicines Adverse Reactions Committee did not collect reports about devices; it took concerted efforts from the women's health advocacy groups to change this. The experiences of doctors and patients remained informal and anecdotal, and, since they were not validated by recognized studies, they were easily outweighed by the formal but unpublished trial information. Many published studies were funded by the population control lobby whose influence was ubiquitous.

The task of instigating change has fallen entirely on consumers. It is the women's health movement that has taken action against these products, and co-ordinated the voices of consumers so that they cannot be ignored. But they have found the regulatory system is weighted against their participation. The first task has been to identify the structural changes that would make the system more evenly balanced. The onus is on individual consumers to demonstrate the existence of a problem with a specific product. Yet, as consumers have found, the regulatory system demands absolute proof that a product is unsafe before it can be removed from the market. In contrast, similar guarantees of safety are not required before a product is allowed onto the market.

References

Anonymous consumer (29 October 1985) Letter to Dr Bob Boyd.
Australian Drug Evaluation Committee (ADEC) (1974) Extract from Minutes of 57th meeting.
Australian Drug Evaluation Committee (ADEC) (28 August 1981) Extract from Minutes of 99th Meeting.
Australian Drug Evaluation Committee (ADEC) (10 December 1982) Extract from Minutes of 106th Meeting.
Brown, H.S. Sales Representative Report for week ending 3 October 1975.
Bunkle, P. (1984) Calling the Shots. In Arditti, R., Klein, R.D. and Minden, S. (eds) *Test Tube Women* 165-87 London, Pandora Press.
Bunkle, P. (1988) A Woman's Right to Choose *Second Opinion: The Politics of Women's Health in New Zealand* 9-27 Auckland, Oxford University Press.
Combly, S., Division of Clinical Services (22 August 1980) Letter to the Sales Manager, G.D. Searle (NZ) Ltd.
Coy, L., Area Medical Research Director Asia/Pacific Region, G.D. Searle (NZ) Ltd (n.d.). Letter to the Director-General of Health.
Dann, C. (1985) Fertility *Up From Under* 51-64 Wellington, Allen and Unwin/Port Nicholson Press.
Edelman, D., Berger, G. and Keith, L. (1982) The use of IUDs and their relationship to pelvic inflammatory disease: A review of epidemiological and clinical studies *Current Problems in Obstetrics and Gynaecology* 6: 1-62.
Else, A. (1991) *A Question of Adoption: Closed Stranger Adoption in New Zealand 1944-1974* Wellington, Bridget Williams Books.

Elstein, M. (April 1967) Pelvic Inflammation and the Intrauterine Contraceptive Device *Proceedings of the Royal Society of Medicine* 60 (4): 397.

Fenwick, P. (1980) Fertility, Sexuality and Social Control over Women in New Zealand. In Bunkle, P. and Hughes, B. (eds) *Women in New Zealand Society* 86-7 Sydney, Allen and Unwin.

Fertility and Maternal Health Drugs Advisory Committee (28 October 1982) Minutes of Meeting included in an Internal Searle Memo (8 November 1982) from King, Subject: Cu-7 Small Transcript.

Food and Drug Act (1969) Acknowledgement Deposited 10 December 1979.

G.D. Searle Ltd (1968-1976) Annual Reports.

G.D. Searle Ltd (6 November 1970) Internal Memo, Subject: Crash Program Studies on the 'Copper 7' IUD.

G.D. Searle (9 November 1970) Internal Memo from William Jenkins to PIC Manager.

G.D. Searle Ltd (20 April 1971) Internal Memo to Dr Carney from Dr Gibor, Subject: Copper 7, Blood Levels of Copper.

G.D. Searle Ltd (29 December 1972) New Drug Application 17-408 p. 164.

G.D. Searle Ltd (1 July 1973) Representatives Report.

G.D. Searle Ltd (April 1975) Internal Memo to R.G. McConnell from P.D. Klimstra, Subject: Liability and Responsibility for Safety Data.

G.D. Searle Ltd (9 June 1977) Internal Memo from F. O'Brien to A. Lynch, Subject: Retraction of Cu-7 Strings, Study patients.

G.D. Searle Ltd (14 September 1982) Internal Memo from King.

G.D. Searle Ltd (14 May 1985) Internal Memo to JEM.

Giffith, R., Director of Clinical Services, Department of Health. Letter to the Scientific Research Associate, G.D. Searle (NZ) Ltd.

Hartmann, B. (1986) *The Right to Live: Poverty, Power and Population Control* San Francisco, Institute for Food and Development Policy.

Hasson, H.M., Berger, G.S. and Edelman, D. (1976) Factors affecting intrauterine contraceptive performance: 1 Endometrial Cavity Length *American Journal of Obstetrics and Gynecology* 126 (8): 973-81.

Hilsden, W.J., Managing Director of G.D. Searle (NZ) Ltd (26 April 1982) Letter to Director-General of Health.

James, P., Scientific Research Associate Asia/Pacific Region (5 October 1979) Letter to the Director-General of Health.

Johns Hopkins University (January-February 1983) Sources of Population and Family Planning Assistance *Population Reports* Series J, number 26, 626.

Kasindorf, J. (5 May 1980) The Case Against IUDs: What Your Doctor Never Told You About Infection, Infertility and IUDs *New West* 21-33, 103.

Line Extension Proposal, Cu7-S (small) Intrauterine Copper Contraceptive (July 1975).

McDonnell, K. (ed) (1986) *Adverse Effects: Women and the Pharmaceutical Industry* 170-1 Penang Malaysia, International Organisation of Consumer Unions, Regional Office for Asia and the Pacific.

Mintz, M. (1985) Deceiving Doctors *At Any Cost: Corporate Greed, Women and the Dalkon Shield* 69-88 New York, Pantheon Books.

Mintz, M. (9 August 1988) The Selling of an IUD: Behind the Scenes at G.D. Searle During the Rise and Fall of the Copper 7 *Washington Post Health*.

Mintz, M. (3 October 1989) Anatomy of a Tragedy *Discovery: New York Newsday* 5.

Morepeth Plant (14 March 1979) Letter from Technical Manager to Cabano, Chicago.

New Drug Application 17-408 (29 December 1972) 148, 159.

Nisbet, T.J. for the Director of Clinical Services (6 June 1986, 17 September 1986) Letter to Mr Hilsden, Managing Director of G.D. Searle (NZ) Ltd.

Nisbet, T., Department of Health (11 December 1987, 14 August 1989) Letter to Mr Van de Water.

Nisbet, T., for Action Manager Medicine and Benefits (20 October 1986, 4 October 1988) Letter to Mr Van de Water.

O'Brien, F., Associate Director of Obstetrics and Gynaecology Research, Letter of 28 October 1975.

Pappert, A. (1986) The Rise and Fall of the IUD. In McDonnell, K. (ed) *Adverse Effects: Women and the Pharmaceutical Industry* 168-9 Penang Malaysia, International Organisation of Consumer Unions, Regional Office for Asia and the Pacific.

Product Introduction Committee (1 December 1970) Minutes.

Risely, R., Principal Medical Officer Medicines and Benefits (19 December 1986) Memo to Minister of Health, Subject: Drug Tariff 1984.

Royal College of General Practitioners (1974) *Oral Contraceptives and Health* New York, Pitman.

Scott, R.G. (1968) Critical illnesses and death associated with intrauterine devices *Obstetrics and Gynaecology* 31: 322-7.

Senanayke, P. and Kramer, D. (1980) Contraception and the etiology of pelvic inflammatory disease: New perspectives *American Journal of Obstetrics and Gynecology* 138: 852-60.

Tatum, H. (1972) Intrauterine Contraception *American Journal of Obstetrics and Gynecology* 112 (7): 1002.

Vance, J. (17 September 1974) Letter to Dr Hubert Lonin, Director of International Medical Affairs, Chicago.

Van de Water, W.A., Managing Director, G.D. Searle (NZ) Ltd (9 November 1971, 6 May 1988, 18 August 1989, 21 September 1989) Letter to Dr Nisbet, Department of Health.

Van de Water, W.A., Managing Director, G.D. Searle (NZ) Ltd (20 October 1986, 10 December 1986, 23 June 1988) Letter to the Director-General of Health.

Van de Water, W.A. (19 July 1989) Letter to Medicines and Benefits Unit, Attention IUCD Advisory Committee.

Villers, T., Sales Manager, G.D. Searle (NZ) Ltd (6 June 1980) Letter to the Director-General of Health, Subject: Pharmaceutical Benefits.

Vosburgh, M. (1978) *The New Zealand Family and Social Change: A Trend Analysis* Wellington, Department of Sociology and Social Work, Victoria University.

Westrom, L. (1980) Incidence, prevalence and trends of acute pelvic inflammatory disease and its consequences for industrialised countries *American Journal of Obstetrics and Gynecology* 183: 880-92.

Westrom, L. and Maardh, P. (1984) Current views on the concept of pelvic inflammatory disease *Australia and New Zealand Journal of Obstetrics and Gynaecology* 24: 98-105.

Wilson, J. (8 August 1980) Letter to J. Phillips, Director of the Division of Clinical Services.

Wright, N.H. and Laemmle, P. (1968) Acute Pelvic Inflammatory Disease in an Indigent Population *American Journal of Obstetrics and Gynecology* 101: 979–93.

Zipper, J., Tatum, H., Pastene, L., Medel, M., and Rivera, M. (1969) Metallic copper as an intrauterine contraceptive adjunct to the T device *American Journal of Obstetrics and Gynecology* 105: 1274–8.

A Living Laboratory: The New Zealand Connection in the Marketing of Depo-Provera

Sandra Coney

In 1968, the injectable contraceptive, Depo-Provera, was introduced onto the New Zealand market with no restrictions on its use. In the 25 years since its introduction, few other developed countries have followed suit. Yet in New Zealand the drug quickly acquired a relatively high level of usage which has been matched by few other countries (Population Information Program, Johns Hopkins University, 1983).

The Upjohn Company which manufactures Depo-Provera has been hampered in its marketing strategies by its failure to achieve approval for the drug in the United States, despite determined efforts over two decades. The animal trials required by the US Food and Drug Administration (FDA) threw into doubt the safety of the drug. Trials on humans which disproved these risks would be necessary before Upjohn could resubmit Depo-Provera to the FDA. New Zealand, with its widespread, long-term use of Depo-Provera provides an ideal study population, and several trials have been inaugurated.

In the 1960s and 1970s, procedures for approving and monitoring the safety of prescription drugs were undeveloped in New Zealand and there has been no systematic collection of information on the use of the drug or on any ill-effects accruing from its use. This has allowed the Upjohn Company and other supporters of the drug to use the New Zealand experience as evidence for its benefits and safety. New Zealand has therefore been a critical element in the Upjohn Company's plans to expand sales of the drug, but by the late 1980s, the results of various trials had thrown the value of the New Zealand connection in the global marketing strategy into jeopardy.

Of Mice and Women

Back in the 1960s, Depo-Provera held all the qualities of a potential market leader for the Upjohn Company of Kalamazoo, Michigan. A long-acting, injectable drug, it was not dependent on anything as fallible as human memory. There was nothing to be done on a daily basis, neither was any action required at the time of sex. Patient 'compliance' was absolute, resulting in an almost totally effective product. In addition, because this injectable contraceptive contained the hormone progesterone, the oestrogenic complications of oral contraceptives, such as blood clots, were avoided.

The new product's chemical name was medroxyprogesterone acetate, abbreviated to DMPA in its depot or injectable form. Upjohn christened it Depo-Provera and in 1967 embarked on the process of gaining permission from the FDA to market its product as a contraceptive.

Upjohn hoped for major sales, not just on the local US market, but also in the Third World, where population control programmes were on the lookout for cheap, easily administered, long-acting forms of birth control. One of the FDA's requirements was that the drug be tested on animals. In 1968 Upjohn began the mandatory trials on rhesus monkeys and beagle dogs.

The results of the animal trials were not helpful to Upjohn's plans. The dogs given DMPA developed breast nodules, and in some cases these were malignant. In one study, five out of 20 dogs given a dose equivalent to the human dose developed malignancies; at higher doses 40 per cent of the dogs grew mammary cancers (Dawson Corporation, 1982). The results of the monkey trials were similarly disastrous for Upjohn. Unexpectedly, some of the monkeys developed uterine cancer, a condition not known to occur spontaneously in this species. Nor were the tests with mice reassuring. They had been poorly conducted and no useful information emerged. When mice died, they had either been cannibalized or had decayed before they could be autopsied (Weisz, Ross, and Stolley, 1984).

In the face of these results, the FDA was not prepared to approve Depo-Provera, but without this approval Upjohn could not sell the drug on the US market and, more significantly, it could not supply the US Agency for International Development, which was the conduit to huge foreign sales. In 1978, Upjohn took the unusual step of appealing against the rejection of DMPA by the FDA. It argued that the animal models were inappropriate and that therefore no weight should be placed on the results of the animal trials. It also argued that because DMPA was already being used extensively in some countries, it must be safe. One Upjohn adviser said that:

To get all glued up about two or three monkeys when there are eleven million women who have used this particular agent and there is not to my knowledge a single documented death and there is not a single documented case where Depo-Provera caused malignancy — I don't know how much more data you would want (Branan and Turnley, 1985).

In 1983 the FDA established a Public Board of Inquiry to hear Upjohn's appeal. It consisted of three persons who were chosen because they were acceptable to both Upjohn and the FDA. It was chaired by Dr Judith Weisz, then Professor at Pennsylvania State University and Head of the Reproductive Biology Section at Hershey Medical Centre.

The Board reviewed the evidence, going back to the original data, a process one Board member later said was important because some studies had acquired a reputation which was not supported by the raw material (Sun, 1984). The Board also held public hearings and commissioned a panel of pathologists to review the specimens from the monkeys that had developed uterine tumours.

The FDA itself argued against the approval of Depo-Provera at the hearings. Fletcher Campbell of the FDA's Bureau of Drugs outlined its position:

Animal data is more worrisome for the drug than for any other drug we know of that is to be given to well people (Branan and Turnley, 1985).

Discussing the results from trials on humans he said:

No adequate test shows the safety of Depo-Provera — not one. Upjohn has had ample opportunity to get the data and they haven't done it. Unlike others, I don't think it is because Upjohn has been negligent. I think we can just as easily assume that they can't get the safety data because the drug isn't safe (Branan and Turnley, 1985).

A year later, Dr Weisz released a report which delivered a serious setback to Upjohn's plans. It said that arguments that the beagle was an inappropriate model were based on 'limited, inconclusive evidence' which was 'inadequate to support the assertion that there are fundamental differences between the dog and the human in the mechanism of action of progestogens'. It came to similar conclusions about the rhesus monkey:

The hypothesis that the cancers originated from an endometrial cell type present in the monkeys but not in the human remains to be tested (Weisz, Ross, and Stolley, 1984).

The report summed up by saying that in animal trials:

If a carcinogenic effect is demonstrated in any one species, in particular, in a site that is expected to be affected by a particular drug or hormonal agent (i.e. a known target organ), this is considered to constitute evidence of lack of safety. If a drug is found to produce cancer in more than one species, the strength of the evidence is increased. (Weisz, Ross, and Stolley, 1984)

In the case of DMPA:

Neoplasms were identified in known target organs of progestogens in two of the species, the dog and the monkey. It would require compelling evidence that the human is uniquely different from each of these two species to disregard these findings (Weisz, Ross, and Stolley, 1984).

The Board was also unimpressed with the evidence Upjohn produced for the safety of DMPA from human studies. It examined over 20 studies which had looked at the cancer risk and found that most of the studies were seriously flawed. DMPA, it concluded, had not been proved safe:

The collection of data from the women who have been using DMPA as a contraceptive worldwide over the fifteen years, has been too haphazard and uncoordinated to provide evidence of the nature and quality required to resolve major outstanding questions concerning the drug's long term safety (Weisz, Ross, and Stolley, 1984).

The Board criticized the failure to inaugurate scientifically-valid human studies to examine long-term risks until Depo-Provera had been on the market over 15 years. The first large studies evaluating the risk of breast cancer and cervical cancer in DMPA users were only commenced in the 1980s, despite the fact that both possibilities had emerged 10 years previously from the animal trials and preliminary trials on humans.

The Board declined to give permission to market DMPA, even for limited use on specific populations of subjects (Weisz, Ross, and Stolley, 1984). The ban on Depo-Provera was upheld.

The New Zealand Connection

Across the other side of the world, Upjohn encountered no such resistance from the regulatory authorities in New Zealand. In 1968, the same year it embarked on its animal trials, Upjohn was given permission to market Depo-Provera in New Zealand, with no restrictions. The Department of Health did not ask for nor did it receive any evidence establishing Depo-Provera's safety.

Over 20 years later, Dr Bob Boyd, Manager of Therapeutics in the Department of Health, said that:

In the early days all you had to do was notify the department that you were marketing a drug. There were no assessment procedures. Upjohn was not expected to produce any of the safety and efficacy data that would be needed for acceptance today (Boyd, 1990).

Since 1968, large numbers of New Zealand women have used Depo-Provera, and New Zealand has been continually quoted as having the highest usage rate in the developed world. In 1987, some 74,200 units of Depo-Provera were used in New Zealand, enough for nearly 20,000 women if they used it for the whole year. Previous estimates had put usage at this level since the mid-seventies. In a study of contraceptive usage among New Zealand women, Dr Charlotte Paul of Otago University found that 13 per cent of the women studied had used Depo-Provera at some time, although only one per cent were currently using it (Paul et al., 1988).

Despite its critical relevance to the tens of thousands of past and future users of DMPA in New Zealand, the FDA Board of Inquiry decision to continue the ban on Depo-Provera was barely noted in New Zealand. The media largely ignored it, and on an official level, support for Depo-Provera continued unabated.

Professor Gavin Kellaway of the Auckland Medical School Department of Pharmacology, chairman of the Department of Health regulatory body, the Medicines Adverse Reactions Committee (MARC) since 1976, said that every time Depo-Provera had been turned down by the FDA, MARC had reviewed the situation in New Zealand:

Each time we have asked the company for evidence and we couldn't find any evidence it was dangerous. It doesn't suit everyone, but there was no evidence of increased cancer (Kellaway, 1990).

In the more than 20 years that New Zealand has used Depo-Provera, the country has assumed an increasing importance in Upjohn's international strategy for DMPA. It was the first developed country to allow unrestricted use of DMPA and only a handful of developed countries have followed suit. New Zealand has used DMPA for a very long time and has contributed nearly half a million woman-years of Depo-Provera use to the global picture.

If something untoward was going to happen to Depo-Provera users, the argument went, wouldn't it have been obvious in New Zealand? For a company sometimes short on good news about its product, New Zealand could be quoted as an example of a country which has had an apparently beneficial relationship with the drug, a kind of living laboratory of successful Depo-Provera use.

The Continuing Controversy

Nevertheless, for almost its entire history in New Zealand there has been continual unfavourable comment about the use of the drug and considerable criticism through the media. In 1990, Dr Tony McCallum, part-time medical director of Upjohn in New Zealand, said that the continual association of the word 'controversial' with DMPA was 'an irritation' to the company. Dr McCallum attributed the failure of Depo-Provera to win marketing approval in the US to the pedestrian nature of the FDA. From the company point of view, he said, Depo-Provera 'is a very good product in business terms. They sometimes talk about it as a "cash cow"'. It was 'a very effective and very appropriate agent with very few disadvantages', although 'no one sees an injectable as a first choice. It's good for someone who can't tolerate oestrogens' (McCallum, 1990).

In many settings in New Zealand, however, Depo-Provera is offered as just one of a number of contraceptive choices women can make. Family Planning Association (FPA) clinics are particularly assiduous prescribers of Depo-Provera. Over a six-month period in 1989, nearly 5000 shots of Depo-Provera were given at FPA clinics throughout New Zealand (New Zealand FPA, 1990).

In 1990, Dr Christine Roke, then acting medical director of the FPA for the Northern Region, commented on the Association's policy on DMPA:

It is Family Planning policy to offer Depo as one of a range of choices. Depo is used by a wide variety of clients, from young through to older women and all ethnic groups. We do use it for people thirteen, fourteen or fifteen years. There is no group we don't advise it for (Roke, 1990).

Doctors were employed by the FPA only if they agreed to prescribe the drug:

We ask new doctors how they feel about using Depo-Provera, and we wouldn't employ someone who wouldn't use it, because it's so much part of our system (Roke, 1990).

Dr Roke said that women who used Depo-Provera successfully liked it because there was nothing to remember, there was no 'messing with condoms and diaphragms', and some women enjoyed the fact that menstruation usually dwindled and could disappear. Another reason was that: 'Some women say it's cheaper, they're not using sanitary pads. Women do come up with these intriguing reasons' (Roke, 1990).

In the experience of the FPA, some who discontinued Depo-Provera did so because of bleeding problems such as heavy bleeding and continuous spotting. Some experiencing these drug effects were offered more hormonal treatment, either oestrogen or more progesterone. Nevertheless: 'There is an occasional haemorrhage. If that happens, we step up the progesterone until it is under control' (Roke, 1990). Dr Roke considered a hysterectomy to control DMPA-induced bleeding 'poor management'.

Dr Roke reported other possible complications:

Some people do put on a lot of weight. They are people who usually put on weight easily. Depo-Provera increases the appetite and makes people eat more. It can be hard work getting the weight off. A minority of women report mood changes and loss of libido (Roke, 1990).

The only life-threatening effect of Depo-Provera reported from an FPA clinic was a case of anaphylaxis, an allergic reaction which caused the patient to stop breathing and nearly die. Dr Roke said: 'We had no adrenalin handy. Now all our doctors and nurses can resuscitate'.

Dr Roke said that the FPA was convinced of the safety of Depo-Provera:

We look at it from the point of view that there have been no deaths on Depo-Provera. It has a better safety record than the Pill or IUD [intrauterine device] (Roke, 1990).

The Upjohn Company held a similar view. Dr McCallum said that:

Usage experience hasn't shown any untoward outcome. Any hazards of a drug would usually appear early, for example, with thalidomide they appeared in the first twelve months of use (McCallum, 1990).

Drug Safety Control in New Zealand

In the absence of organized studies, it would in fact be difficult for evidence of any dangers of Depo-Provera to emerge, even from the kind of widespread use found in New Zealand. Cancers are uncommon events and thousands of women would need to be carefully studied to detect any increase in cancers in Depo-Provera users. In addition, cancers grow slowly and years could elapse between use of a drug and any serious adverse effect.

Dr Judith Weisz and her colleagues on the FDA Public Board of Inquiry explained why an apparent lack of problems with a drug was not a reliable means of measuring its safety:

There are few circumstances under which an adverse reaction to a drug will make itself obvious in the absence of systematic study (Weisz, Ross, and Stolley, 1984).

Good information, the Board said, cannot be obtained 'from casual observations' (Weisz, Ross, and Stolley, 1984). Besides this, it would be difficult for evidence of serious risks with Depo-Provera to emerge from New Zealand's process for reporting adverse drug effects. Doctors do not reliably report ill-effects of drugs, and are not always capable of identifying the symptoms the patient presents with as drug effects.

The New Zealand system for monitoring drug reactions is not well developed. The Medicines Adverse Reactions Committee (MARC), the committee which monitors drug safety in New Zealand, was started as a voluntary committee by the Otago Medical School in 1964. Until 1972, members even paid their own airfares to attend meetings in Wellington. Only in the 1980s did the Department of Health start paying fees to members, and the system is still under-resourced.

Only 101 adverse reactions in women using DMPA have been reported to MARC in over 20 years of use (MARC, 1990). These range from relatively minor problems such as acne and headaches (and one 'increase in libido') to serious reactions such as nine reports of anaphylaxis and several cases of pulmonary embolism. Whether any of these reactions can definitely be related to DMPA use is not established by the reporting system.

According to Professor Gavin Kellaway, doctors do not report all adverse reactions with a product:

They are asked to report severe or unexpected reactions, not minor ones, nor ones that are common or recognized. For instance, sexual problems or depression would be unlikely to come our way. Voluntary reporting is not a good indicator of how often something occurs. We use the term 'suspected reaction' because we can't check back to see if it really was a drug reaction (Kellaway, 1990).

One method of ensuring more complete collection of adverse effects of a drug would be to put Depo-Provera onto a system called 'monitored release'. This would require Upjohn and doctors to report every event and side-effect associated with use of the drug. MARC has only previously put one drug on monitored release after it had been freely listed, the antidepressant Mianserin, which under the voluntary reporting system turned up unexpected cases of depression of white blood cells and deaths. Under monitored release, even more cases emerged and the drug was taken off the free Drug Tariff, restricting its use.

According to Professor Kellaway:

The advantage of monitored release is that you get a lot of reports of things that may seem irrelevant, but which turn out to be significant. For example, you would have to report events, even fractures. Now it may seem unlikely that Depo-Provera would cause fractures, but if you got a lot of reports of that, you'd have to wonder if it did something to bones (Kellaway, 1990).

At the FDA Board of Inquiry hearings, one Upjohn consultant said that:

There is no perfect animal model and in a sense the final animal model for a drug to be used on humans, has to be the human (Branan and Turnley, 1985).

That is precisely what New Zealand with its long and widespread Depo-Provera use is able to provide medical science. The FDA Public Board of Inquiry commented that New Zealand was ideal for studies because of its long-term Depo-Provera use.

Adequate data collection should be feasible and the incidence of neoplasias (cancers) of the breast and endometrium are comparable to those in the United States (Weisz, Ross, and Stolley, 1984).

Depo-Provera and Bone Density

Several studies have been initiated in New Zealand since 1979. The largest, the Upjohn-funded New Zealand Contraception and Health Study (NZCHS), is due to be completed in the early 1990s. In 1989, a study by Dr Charlotte Paul of breast cancer risk in Depo-Provera users was published in the *British Medical Journal* (Paul, Skegg, and Spears, 1989). Both these studies are concerned with the risk of cancer in users of DMPA. For all the decades of its use, the possibility that Depo-Provera might cause cancer has been the overwhelming focus of attention and study. But in 1991, an Auckland study raised a new and unexpected issue: the curious problem of whether Depo-Provera could adversely affect bones.

The possibility of loss of bone density in Depo-Provera users had been raised at the FDA's Public Board of Inquiry and its report had called for these concerns to be 'resolved promptly by appropriately designed studies' (Weisz, Ross, and Stolley, 1984). But despite this, no such studies were carried out. The issue was overlooked.

In New Zealand, Dr Tim Cundy, an endocrinologist in the Department of Medicine at Auckland Hospital, was accidentally alerted to the possibility that Depo-Provera might exercise a detrimental effect on bone density. His clients were elderly women with osteoporosis, a condition caused by thin bones which can lead to fractures. Older women are prone to develop osteoporosis in part because their ovaries stop producing oestrogen at menopause, and oestrogen has a protective effect on bones.

Dr Cundy became interested in Depo-Provera's effect on bones after the daughter of a woman attending his clinic asked what her risk of developing the condition was. Her oestrogen levels were measured, and were found to be unexpectedly very low. Dr Cundy said: 'We were all scratching our heads, wondering what had caused it. Then we found she was on Depo-Provera' (Cundy, 1990).

Depo-Provera switches off the pituitary, the gland at the base of the brain which secretes various hormones which control what happens in the ovaries. Thus, in addition to its action in preventing ovulation, Depo-Provera reduces oestrogen levels.

Working with Dr Helen Roberts, medical director of the Auckland FPA, Dr Cundy recruited 30 women under the age of 45 years each of whom had been using Depo-Provera for over five years. These women were matched to a control group of non-Depo-Provera users for age, race, and body size. The study compared the two groups for their hormone levels, and bone density in the hip and spine was measured on a dual energy x-ray absorptiometry machine. Depo-Provera users were also compared with a group of normal postmenopausal women who would be expected to show bone loss.

The results showed that the young women using Depo-Provera had a significant reduction in bone density (Cundy et al., 1991). On average, the women using Depo-Provera had 7.5 per cent less bone in the lower spine and 6.5 per cent less in the hip than women who had never used Depo-Provera. In the group of normal postmenopausal women, bone density was reduced by 16 per cent in the spine and 10 per cent in the hip. These results gave young DMPA users bone-density values intermediate between those of normal premenopausal and postmenopausal controls. The study concluded that:

> Long term use of DMPA is associated with significant reductions in bone density in the lumbar spine and femoral neck. Use of DMPA should therefore be considered a potential risk for osteoporosis (Cundy et al., 1991).

Dr Cundy described these results as 'worrying'. He said this level of loss in the DMPA users:

> would increase their risk of osteoporosis by 50 per cent. It would increase the lifetime risk of fracture. If you start that much lower on the scale, your risk may be increased. It would differ according to the kind of bones you had. People who start off with good bones could probably lose some bone density without ill effect, but people with little bones, could ill afford to lose the amount they do on Depo-Provera (Cundy, 1990).

Dr Cundy described his results as statistically 'quite secure, although it would be gratifying if someone else came up with the same result' (Cundy, 1990). It is possible that women could regain bone density when they ceased using Depo-Provera, as women do after using other drugs which cause bone loss, such as some used in the treatment of endometriosis.

There are also particular groups of women who might face additional problems. Women who use Depo-Provera right up to menopause would have no chance to regain bone density before oestrogen production in their ovaries started to naturally slow down, while women who started using Depo-Provera immediately after childbirth might be at special risk, because bone is normally lost during pregnancy and breast feeding.

Dr Charlotte Paul explained why the results of the Auckland Hospital study were not surprising:

The results are biologically plausible. We know that athletes who stop having periods also have lower bone density and it also fits with what is known of the role of oestrogen in bone loss (Paul, 1990).

Dr Christine Roke was similarly 'not surprised'. She had heard the subject of bone density loss discussed at an International Planned Parenthood Conference in Thailand, and within her work she had 'occasionally come across someone grossly oestrogen-deprived, vaginally, on Depo-Provera' (Roke, 1990).

Dr Ruth Bonita, Masonic Senior Research Fellow in the Geriatric Unit at the University of Auckland, said that if Dr Cundy's results are replicated by further studies it would have 'grave repercussions' (Bonita, 1990). Osteoporosis is a growing problem in western countries as more women live longer; it is possible that the problem might be aggravated in New Zealand because of its widespread use of Depo-Provera. According to Dr Bonita:

It is difficult to know how big a public health problem that would be. It's another example of the way in which the medicalisation of women's lives has detrimental iatrogenic effects (Bonita, 1990).

Dr Cundy received funding from Upjohn for the biochemical tests needed to carry out his research. On the completion of the study, Dr Cundy informed Upjohn of the results, and according to him, 'they were not terribly pleased' (Cundy, 1990).

Following this study, Dr Cundy sent Upjohn the protocol for a new study to examine the bone density of women who had stopped using Depo-Provera, and applied for funding. He was visited by the company's international medical director, Dr David Zambrano, from Argentina, who took the protocol to the Upjohn Company headquarters in Kalamazoo. It was returned with 'ten pages of quibbles' (Cundy, 1990), and funding was declined.

Dr Cundy said his work did not deliver:

a killing blow to Depo-Provera. The results need to be confirmed by another study. But women should be informed about the risk if starting Depo-Provera. We should probably suggest that if they have other risk factors for osteoporosis, they shouldn't start Depo-Provera (Cundy, 1990).

Dr Tony McCallum is cautious about the implications for Upjohn. He described the research as 'a curiosity-type study': 'You may have a finding, but there may be more than one way of explaining it' (McCallum, 1990).

The Issue of Breast Cancer

But if Upjohn remained phlegmatic about Dr Cundy's research, it reacted strongly to the results of Dr Charlotte Paul's research into the association between breast cancer and Depo-Provera use. Dr Paul's study explored the question of the relevance of the trials on beagle dogs for human users of DMPA. In its attempts to gain approval for DMPA, Upjohn has purported to clear the drug of any imputation of causing a risk of breast cancer in DMPA users (Minkin, 1981). The large, well-conducted New Zealand trial posed a potential threat to Upjohn's claims that 'no direct link between Depo-Provera and breast cancer in women has ever been established . . .' (Upjohn Company, 1986).

Breast cancer is a significant cause of death among New Zealand women, as it is among women anywhere in the world. Any increase in risk would have serious public health consequences, and unlike cervical cancer, there is no screening test which can detect cell changes before the cancer occurs.

Dr Paul and her co-authors at Otago Medical School looked at the contraceptive histories of all women in New Zealand who had had breast cancer diagnosed between 1983 and 1987. While they found no overall increased risk from Depo-Provera use, when they looked at sub-groups, they discovered that women who had had breast cancer diagnosed at a young age were twice as likely to have used the drug. Similarly, use before the age of 25 years and use for a long period increased the risk (Paul, Skegg, and Spears, 1989).

Publicly, Upjohn hardly reacted to this, but when MARC discussed Dr Paul's study in late 1989 to determine what, if any, action it should advise the Health Department to take, Upjohn lobbied the Committee strenuously. According to the Health Department's Manager of Therapeutics, Dr Bob Boyd, this kind of lobbying by drug companies was not new: 'They have a right to advise us, but I think approaching each member of the advisory committee is unusual' (Boyd, 1990).

In any event, MARC decided to wait on the results of another study looking at breast cancer and Depo-Provera to see if it confirmed Dr Paul's findings. This is a large World Health Organization (WHO) hospital-based multinational case-control study which has been running for some years. The Department of Health's only action was a brief summary of Dr Paul's results in its March 1990 *Clinical Services Letter* sent to all general practitioners with the advice that no changes were yet warranted in doctors' prescribing habits (Department of Health, 1990).

Dr Boyd described this as 'a type of warning and an intimation that something else was coming, but the company would rather it hadn't been done' (Boyd, 1990).

Critics of Depo-Provera might have been annoyed at the timidity of this response, but Upjohn was not pleased. Dr Zambrano, Dr McCallum, and other Upjohn officials wrote to the *New Zealand Medical Journal* concerned that what they called Dr Paul's 'speculations' might be 'misinterpreted' (Day et al., 1990). They criticized the statistical analysis in the paper and said the positive results for young users were a 'chance relationship'.

Dr Tony McCallum said that the company regarded the Department of Health's *Clinical Services Letter* as 'irresponsible': 'We were a little cross. It's statistically a poor trial, it's very suspect and that's the position the company takes' (McCallum, 1990).

But Dr Charlotte Paul and her co-authors maintained that Upjohn has not understood the paper:

Theirs are simple mistakes, springing from a lack of basic epidemiological knowledge, which could have been corrected by seeking advice from us or other epidemiologists (Paul, Skegg, and Spears, 1990).

Dr Paul said the study is 'considered to be one of the major studies contributing to knowledge internationally'. The WHO study 'is not better, it's just much more powerful if two studies say the same thing' (Paul, 1990).

Most previous human studies of breast cancer and Depo-Provera had had design problems which meant that it was not possible to give much weight to their results. Defects in these studies were comprehensively listed by the FDA Public Board of Inquiry (Weisz, Ross, and Stolley, 1984). But a later study, in 1987, which received little attention, had also shown an increased breast cancer risk in DMPA users (Lee et al., 1987). The study was conducted in Costa Rica, a country with a high level of Depo-Provera use. Unlike Dr Paul's study, this study showed that Depo-Provera use had 2.6 times the normal risk of breast cancer. According to Dr Paul:

> Although the study involved small numbers of Depo-Provera users, this was a statistically significant finding, but the authors themselves worried about why they were getting something different from the WHO results. They had only interviewed 67 per cent of the relevant breast cancer cases. The rest had died and it is possible that those who had died were less likely to be Depo-Provera users (Paul, 1990).

An initial report from the WHO study showed no overall risk (WHO, 1986), but unlike Dr Paul's study, the study had not looked at sub-groups. According to Dr Paul, the lack of an overall risk was:

> completely consistent with our findings. What we're waiting on are the results for the sub-groups, for the young women, to see if their results are the same as ours (Paul, 1990).

Dr Paul's study caused some change of policy at FPA clinics. Dr Christine Roke described herself as 'a bit staggered' at the result:

> But it does fit in very well with what's coming out on the Pill, that there's an association between Pill use by young women and risk of breast cancer (Roke, 1990).

She believed that this risk would probably be proven by later studies. In response to the study's findings, a decision was made to inform women under 25 attending FPA clinics about these possible risks.

Possible Effects on the Cervix

The third study important for the future of Depo-Provera is the New Zealand Contraception and Health Study (NZCHS) (NZCHS Group, 1986). This study, involving 7500 women, 3500 of whom are using Depo-Provera, is principally looking at the relationship between Depo-Provera and cervical cancer. The WHO study on breast cancer risk was also studying DMPA's relationship to cervical cancer and the preliminary results did show an increased risk, but only in women who used the drug for more than five years and who were under 30 when they first used it (WHO Collaborative Study of Neoplasia and Steroid Contraceptives, 1985). This association was described as tentative. The NZCHS is seen as a major investigation of DMPA's effect on the cervix. It has been continuously surrounded by controversy since it was announced in 1979.

Early criticisms from statisticians were that the number of study participants was too small and the length of the study too short to prove or disprove whether Depo-Provera caused cervical cancer (Renner, 1983; Cryer, (1984); Hassard, (1984); and Renner, 1984). An internal memo from Upjohn dated 1977 fuelled suspicions that the study was destined to produce a favourable result for Upjohn (Mohberg et al., 1977). This memo to company officials which reviewed an initial study proposal said that the study had the advantage that:

The chances of results strongly unfavourable toward Depo-Provera are very low, because of the low carcinoma-in-situ rates among New Zealand whites (Mohberg et al., 1977).

This memo also estimated that the sample size was half as big as it needed to be to detect an increase in cervical dysplasias (abnormalities) in users of DMPA and that this would lead to a lack of credibility in the study:

The critics will very quickly recognize that the proposed study is powerless to detect any but the most extreme differences (Mohberg et al., 1977).

The lack of power in the study was vigorously disputed by members of the NZCHS Executive Committee (Remington, 1984) and there was confusion about the end-point of the study, whether it was carcinoma-in-situ or earlier stages of dysplasia. The initial proposal underwent some changes, and recent data, which has shown a much higher rate of dysplasias among New Zealand women than previously realized, added weight to the view that the study *did* have sufficient power to produce a valid result (NZCHS, 1989).

There have also been predictions that it will be difficult to unravel which contraceptive is the culprit if a woman develops a cervical dysplasia. This aspect was called by the NZCHS Executive Committee 'statistically challenging' (Remington, n.d.). FPA clinics have been closely involved in recruiting women for the study, and Dr Christine Roke outlined the problem:

It is an observational study so women use what they want. When they come into the study, they are using one form of contraception, but they chop and change a lot. They have babies, have their tubes tied, or their husbands have vasectomies. I would suspect almost a minority stay on the same contraceptive right through (Roke, 1990).

These conditions make the quality of the data analysis of crucial importance, and it is this aspect of the NZCHS which has been an area of major contention.

The argument concerns the independence of the study. The NZCHS is funded by Upjohn (to the massive sum of $6 million), the study headquarters are at the Upjohn Company premises in Auckland, the data is stored on the company computer at Kalamazoo, and this is also where the analysis is taking place (NZCHS, 1981; Bunkle, 1984).

The principal researcher is Professor Sir G.C. Liggins, a New Zealand researcher of international repute whose major field has been prostaglandins. The Executive Committee in charge of the study is composed of some of the most eminent statisticians and scientists in the world, although there is also an Upjohn Company representative on it. Criticisms remain that the analysis of the data is insufficiently removed from the Upjohn Company.

It was for this reason that in the 1970s Dr Robert Beaglehole, now Professor and Head of Auckland Medical School's Department of Community Health, declined to be involved in the study:

I was approached by Upjohn and flown to Kalamazoo to discuss my involvement at the New Zealand end. I was extremely unhappy with the involvement of the company in the design, execution, analysis, and publication of the study. It seemed inappropriate to me that the company should be so intimately involved with evaluating its own product. Other companies at the time had been criticized for the same thing. The company should be several steps removed from the study. As it is, the study is funded, serviced, and supported directly by the company (Beaglehole, 1990).

Professor Beaglehole was also concerned that the study had never been submitted formally to a New Zealand ethics committee for approval. Because the study had been funded by Upjohn, and was being conducted through general practitioners and FPA clinics, there was no obligation for it to be submitted for ethical approval.

In the aftermath of the Cartwright Inquiry and the exposure of unethical research practices at National Women's Hospital in Auckland (Committee of Inquiry, 1988), Professor Liggins asked the Medical Research Council (MRC) to consider the ethics of the NZCHS in 1989.

The MRC's findings, released late in 1989, stated that it was not possible to give 'retrospective ethical consent' to the study (MRC, 1989). The Committee said that ethical standards had changed since the study was planned, and it recommended 'that this type of study never again proceed . . . without review by a properly constituted ethical committee' (MRC, 1989). Because of the time the study had been running, the Committee felt it would be unethical to stop it, but it strongly recommended that 'a duplicate system of peer scientific review and statistical analysis' be established to provide an independent review of the analysis of the data. This should be based in New Zealand, it said, and the cost should be borne by Upjohn.

Professor Beaglehole, who made a submission to the review, thought this 'an excellent idea. I think Mont Liggins should bend over backwards to see it happens' (Beaglehole, 1990).

Initially, the MRC was hopeful that the matter might be satisfactorily resolved. In early 1990, the secretary of the MRC at the time, Jim Borrows, said:

We had tried to find an alternative analytical pathway. Because of the nature of this study and the fact that it deals with local data, it would be wise for Professor Liggins to press the study funders to pay for independent analysis. When people undertake research, they have obligations to see it is done properly. We're throwing the responsibility back to him. Without an independent analysis, there will inevitably be criticism when the study is published. He may be prepared to stand that, that's his business, and the criticism may be unjustified, but the way to deal with it is to get rid of it (Borrows, 1990).

Professor Liggins indicated to the MRC that he was open to a New Zealand panel of biostatisticians analysing the data. However, there was disagreement over who should pay. Professor Liggins' view was that:

It's not our responsibility to do so. If others wish to reanalyse the results, they're welcome to do so, but I see no reason why we should have to pay for it (Liggins, 1990).

In August 1990, Professor Liggins responded formally to the MRC. The letter was brief and simply stated that the Executive Committee of the NZCHS had looked at the MRC review report and its 'contents were noted' (Hodge, 1990), but nothing further was to be done.

Dr Jim Hodge, director of the MRC at the time, expressed disappointment at this response:

It is really a negative reply. They are not prepared to take that extra step of covering themselves by adopting the recommendations of the MRC (Hodge, 1990).

Professor Liggins was adamant that as it stands the integrity of the study was being maintained, despite the close links with the Upjohn Company. An independent audit of the data by Dr Morton Hawkins from the University of Houston's School of Biostatistics was incorporated in the protocol, and in defence of the study process, Professor Liggins cites his own reputation:

> I can see no reason to risk what standing I have in international and local scientific communities by being part of a study that failed to observe stringent standards of scientific publication. When the results come out, I have no doubt some will want to cast doubt on the validity of the findings — unless they are adverse in which case they will be silent (Liggins, 1990).

Nevertheless, there is a good deal of scepticism within New Zealand and internationally about the study. Judy Norsigian, chair of the Depo-Provera Committee of the US National Women's Health Network, the major women's lobby to put the consumer viewpoint to the FDA Board of Inquiry, said that internationally:

> The NZCHS is looked at quite warily because of the way the data is being analysed. It was always a mistake not to leave the analysis to the New Zealand scientists. I don't think whatever the result is that there will be uniform recognition (Norsigian, 1990).

The Enduring Uncertainty

After nearly a quarter of a century's use on human beings, including tens of thousands of New Zealanders, the issue of whether Depo-Provera is safe or not is still unresolved. The Otago University study has raised the possibility of a breast cancer risk to some groups of users, the Auckland Hospital study suggests that Depo-Provera causes loss of bone density, and the results of the NZCHS examining the relationship between the drug and cervical cancer are not yet available.

The question of the drug's safety has been repeatedly raised by MARC. According to Dr Bob Boyd: 'Depo-Provera has been before MARC more than most medicines' (Boyd, 1990). Dr Gavin Kellaway said that:

> If the results of the study are highly suspicious, then that would be the kind of indication that would lead to DMPA being put on monitored release or even taking it off altogether. It depends how firm the data is (Kellaway, 1990).

In its history, the New Zealand Department of Health has rarely removed a drug on the free list from the market for reasons of safety. In 1991, it did take steps to remove certain doses of the sleeping tablet, Halcion, another Upjohn Company product, after reports that it could induce psychiatric disturbances as a side-effect. However, in taking this uncustomary action, the New Zealand Department of Health was able to follow the action of its British counterpart which had previously taken

Halcion off the market. (The Associate Minister of Health later suspended the action of the New Zealand Department of Health.)

Because Depo-Provera is not freely used in other developed countries, no other country is likely to provide a lead for the New Zealand drug regulatory authorities. On the contrary, in the case of Depo-Provera, other countries might look to New Zealand for guidance on the drug's safety performance.

The New Zealand Department of Health's reluctance to take stringent action over unsafe drugs has led it to adopt other strategies to minimize their use. One tactic is to take a drug off the Drug Tariff so that it is no longer free of charge. This was the strategy used by the Department with Mianserin, and with G.D. Searle's Copper 7 intrauterine device after it was withdrawn from the US market but not the local one. In other cases, such as the Lippes Loop IUD, the Department has put pressure on the manufacturer or distributor to voluntarily withdraw the product. According to Dr Boyd:

Removing DMPA from the Drug Tariff would make use go right down as it would be reserved for people for whom it's really needed, but this is only a control measure. It doesn't guarantee safety for people. If we're talking about cancer-causing potential, then I'm not sure that's sufficient (Boyd, 1990).

Dr Tony McCallum said that while his company was confident about the drug, if the NZCHS showed:

an unfavourable result, Depo-Provera would be recalled one way or another. It would be a surprise because general clinical use wouldn't indicate such a tendency. All the company can do is remain vigilant and if any finding shows a definite association, where the drawbacks outweigh the benefits, you can restrict users. Obviously if an incontrovertable hazard happens, the company does step in. There would be tremendous shock waves if the company's confidence in Depo-Provera is found to be grossly misplaced (McCallum, 1990).

A restriction or ban on DMPA in New Zealand raises the question of what implications this would have for Depo-Provera internationally. Dr Tony McCallum painted a picture of the multinational Upjohn as comprising relatively autonomous units:

Each subsidiary looks after its own world. We feel a bit out on a limb in New Zealand. When we jump, the rest of the world doesn't jump (McCallum, 1990).

He said that Upjohn had abandoned hope of ever getting approval to market DMPA in America:

I gather the company just says that's bad luck. If US society wants to live without Depo-Provera, it can live like that. But in New Zealand, many doctors and patients are satisfied. We guard it jealously (McCallum, 1990).

But Dr Boyd argued that the international connections were strong:

Anything that happens in a country where they've had approval for a while would have international repercussions. So I'm sure they've had instructions from the parent company to defend Depo-Provera in New Zealand. To defend their existing markets against other companies, they will try and preserve the New Zealand market (Boyd, 1990).

Dr Boyd said that were New Zealand to take action against Depo-Provera, regulatory agencies in other countries and the WHO would know:

They would sit up and take notice. What we do is accepted as the action of a developed country with repercussions for international markets. If it was done by Senegal no one would take much notice. Depo-Provera is very important to WHO's contraceptive programme and to the company's reputation. I wouldn't feel too secure if I was their product manager with Halcion and Depo-Provera and not too many new products coming forward (Boyd, 1990).

Upjohn faces other threats from competitors to its international marketing prospects. In the many years Upjohn has been trying to prove its safety, Depo-Provera has run out of patent. In addition, other newer progesterone products may have overtaken Depo-Provera. The main contender is Wyeth's Norplant, progesterone-releasing implants which are sewn into the flesh of the forearm and render a woman sterile for five years at a time.

According to Judy Norsigian of the US National Women's Health Network:

Norplant's the progestin approach that is getting the most attention in America. We don't get the sense that Depo's going anywhere. They can't prove its safety and people are moving onto other things. Upjohn has lost its monopoly and I think they take the attitude, why fight this thing now? The short-term data for Norplant is not terrible. Although the women's health groups opposed approval for Norplant and want more research, the problems with Norplant are not so severe as to say it shouldn't be on the market. It is reversible and has less intense effects on the body and the cancer problems didn't come up (Norsigian, 1990).

Norplant was approved for contraceptive use by the FDA in late 1990. According to Judy Norsigian, 'the major issue is how to have the implants removed quickly and skilfully when women want it'. Cases have been reported where doctors refuse to remove it: 'We worry women might resort to self-removal' (Norsigian, 1990).

Conclusion

For nearly 25 years, New Zealand women have had Depo-Provera offered to them as a first choice contraceptive with no restrictions on use, and about 4 per cent of married women of reproductive age (Population Information Program, Johns Hopkins University, 1983) and an unknown number of other women have been using it at any one time. Despite the fact that the drug underwent no scrutiny before entering the New Zealand market, and despite numerous unresolved safety questions, the New Zealand Department of Health has never placed any restrictions on its use, nor has it warned doctors of risks. The only information available to women is provided by the Upjohn Company and by groups which support its unrestricted use.

The Upjohn Company has vigilantly nurtured this unrestricted New Zealand market. As the major developed country with widespread Depo-Provera use, New Zealand is valuable to the Upjohn Company. The fact that the Department of Health has never restricted its use could suggest to those outside the country that no problems have arisen. However, limitations in the New Zealand system of collecting drug adverse effects and the history of timidity on the part of the Department in its drug regulation behaviour make it very unlikely that any action would be taken.

New Zealand is also important to the Upjohn Company because it can provide a location for studies which will gain credibility through the calibre of the researchers, the universal health care coverage available to New Zealanders, and the racial composition of the population.

On the other hand, these factors have also presented problems to Upjohn. In the past five years independent research has been carried out that raises rather than dispels concerns about Depo-Provera's safety. Upjohn has done its best to minimize the adverse effects of such findings and to maintain Depo-Provera's availability on the market. Its tactics have involved criticizing the quality of research which has unpalatable findings, lobbying the Department of Health and its committees, and applying pressure through the media.

The history of this chapter provides an example of the latter strategy. It was originally published in a shortened version in the national magazine *The Listener* (Coney, 1990). During the period of its preparation, the Upjohn Company wrote to the magazine's editor criticizing the author, and offering the services of medical experts to 'vet' the article. A package of pro-Depo-Provera papers was also enclosed. This tactic resulted in the submission of the article for legal scrutiny by *The Listener* before it was published.

Postscript

The long-awaited WHO study on breast cancer was finally published in October 1991 (WHO Collaborative Study of Neoplasia and Steroid Contraceptives, 1991). It essentially confirmed the findings of the study conducted by Dr Paul. The study compared young women with breast cancer admitted to hospitals in Kenya, Mexico, and Thailand with women of a similar age admitted for other conditions. Overall, the breast cancer cases were no more likely to have been Depo-Provera users than the controls. However, when sub-groups were examined, an increased risk was found for the cases in some categories. These were women who were recent users. Risk was increased within the first four years of initial exposure, mainly in women who were under 35 years of age. In these women the risk of breast cancer was more than doubled among DMPA users.

Women who had used DMPA more than five years previously were not at additional risk, and the risk did not increase with duration of use.

The immediate public reaction to this study within New Zealand continued the previously established pattern of silence from Upjohn and manipulation of the meaning of the results to the advantage of Depo-Provera from proponents of DMPA.

The study results were released to the media by the press secretary of Contraceptive Choice, a group which supports the unrestricted use of Depo-Provera, several weeks before the study itself was published (Barnes, 1991). The press who picked up the media release did not have access to the study to interpret it for themselves. Instead, they took their lead from the media release.

The release was headed 'Depo Provera should continue to be used as a contraceptive' and said the study results were 'reassuring'. The risk of breast cancer in the users of DMPA, it said, was 'no greater than with contraceptive Pill'.

The media adopted this line and failed to comprehend the significance of the study in reinforcing the earlier New Zealand study.

The major television news coverage said that Depo-Provera was as safe as the Pill, and a New Zealand FPA spokesperson said that the Association would never have used the injection if there had been any question as to its safety.

During an earlier investigation which forms the basis of this chapter, the author had been told by the FPA that confirmation of the New Zealand findings on breast cancer risk in DMPA users by the WHO study would lead to re-evaluation of its use. However, when the study results were essentially confirmed by the WHO study, the Association continued to advocate for Depo-Provera, and turned the WHO results into 'good news' for the drug. It was now 'as safe as the Pill'. There was no public response from Upjohn. In November 1991, on the advice of MARC, the New Zealand Department of Health decided to take no action over Depo-Provera.

References

Barnes, L. (17 September 1991) Depo Provera (press release).
Beaglehole, R. (20 April 1990) Interview with the author.
Borrows, J. (3 May 1990) Interview with author.
Boyd, B. (23 August 1990) Interview with the author.
Branan, K. and Turnley, B. (1985) The Ultimate Test Animal (video), USA.
Bunkle, P. (1984) Calling the shots? The international politics of Depo-Provera. In Arditti, R., Klein, R.D. and Minden, S. (eds) *Test Tube Women* 165-87 London, Pandora Press.
Committee of Inquiry in Allegations Concerning the Treatment of Cervical Cancer at National Women's Hospital and Other Related Matters (1988) *The Report of the Cervical Cancer Inquiry* Government Printing Office, Auckland.
Coney, S. (12 November 1990) Women at risk *The Listener* 10-15.
Cryer, C. (1984) Comment on 'Depo-Provera: The New Zealand Contraception and Health Study' *The New Zealand Statistician* 19 (1): 20-4.
Cundy, T. (7 May 1990) Interview with the author.
Cundy, T., Evans, M., Roberts, H., Wattie, D., Ames, R., and Reid, I.R. (1991) Bone density in women receiving depot medroxyprogesterone acetate for contraception *British Medical Journal* 303: 13-16.
Day, B.J., McCallum, A.B., Zambrano, D., and Tong, D.D.M. (1990) Depot medroxyprogesterone and breast cancer (letter) *New Zealand Medical Journal* 103: 327.
Dawson Corporation (1982) Long-term Depo Provera Study in Dogs, Interim Findings, and also Final Report, June. DMB Vol Nos 255-73, Tab 295, cited in Weisz, Ross, and Stolley, 1984.
Deely, J.J., Convenor, Survey Appraisals Committee of New Zealand Statistical Association (1983) The New Zealand Contraception and Health Study: the Appraisal of the Protocol *The New Zealand Statistician* 18 (2): 6-11.
Department of Health, New Zealand (15 March 1990) Depo-Provera and Breast Cancer *Clinical Services Letter* 257.
Department of Health (1990) Statistical information on Depo-Provera use through prescription and practitioners' supply orders, communication to the author.
Lee, N.C., Rosero-Bixby, L., Oberle, M.W., Grimaldo, C., Whatley, A.S., and Rovira, E.Z. (1987) A case-control study of breast cancer and hormonal contraception in Costa Rica *Journal of the National Cancer Institute* 79: 1247-54.
Hassard, T.H. (1984) Comment: Problems in the design of cancer trials: some thoughts on the Depo-Provera controversy *The New Zealand Statistician* 19 (1): 25-8.

Hodge, J. (1990) Communication to the author.

Kellaway, G. (17 April and 11 May 1990) Interviews with the author.

Liggins, M. (26 April 1990) Interview with the author.

McCallum, A.B. (21 May 1990) Interview with the author.

Medical Research Council (MRC) (1989) Summary of a Review by the Medical Research Council's Committee on Ethics in Research of the New Zealand Contraception and Health Study (unpublished).

Medicines Adverse Reactions Committee (MARC) (May 1990) Print-out of reactions to Depo-Provera, communication to the author (unpublished).

Minkin, S. (November 1981) Nine Thai women had cancer . . . None of them took Depo-Provera: Therefore Depo-Provera is safe . . . This is science? *Mother Jones* 34–50.

Mohberg, N.R., Assenzo, J.R., and Schwallie, P.C. (1 November 1977) Memo to D. Weisblat, Subject: Review of Depo-Provera study proposal (unpublished).

New Zealand Contraception and Health Study (NZCHS) (May 1981) Summary (unpublished).

New Zealand Contraception and Health Study (NZCHS) Group (1986) New Zealand contraception and health study: design and preliminary report *New Zealand Medical Journal* 99: 283–6.

New Zealand Contraception and Health Study (NZCHS) Group (1989) The prevalence of abnormal cervical cytology in a group of New Zealand women using contraception: a preliminary report *New Zealand Medical Journal* 102: 369–71.

New Zealand Family Planning Association (FPA) (1990) Clinic Statistics for the six months July–December 1989 (unpublished).

Norsigian, J. (19 May 1990) Interview with the author.

Paul, C., Skegg, D.C.G., Smeijers, J., and Spears, G.F.S. (1988) Contraceptive practice in New Zealand *New Zealand Medical Journal* 101: 809–13.

Paul, C., Skegg, D.C.G., and Spears, G.F.S. (1989) Depot medroxyprogesterone (Depo-Provera) and risk of breast cancer *British Medical Journal* 299: 759–62.

Paul, C. (12 May 1990) Interview with the author.

Paul, C., Skegg, D.C.G., and Spears, G.F.S. (1990) Depot medroxyprogesterone and breast cancer (letter) *New Zealand Medical Journal* 103: 327–8.

Pool, I. (December 1980) The Depo-Provera controversy *New Zealand Population Review* 6 (4): 26–36.

Population Information Program, Johns Hopkins University (May 1983) Long-Acting Progestins — Promise and Prospects, Injectables and Implants *Population Reports* Series K, Number 2.

Population Information Program, Johns Hopkins University (March–April 1987) Hormonal Contraception: New Long-Acting Methods, Injectables and Implants *Population Reports* Series K, Number 3.

Remington, R.D. (1984) Comment *The New Zealand Statistician* 19 (1): 29–30.

Remington, R.D. (no date) Comments by Professor Remington on Mr Renner's manuscript (unpublished).

Renner, R.M. (1983) Depo-Provera: the New Zealand Contraception and Health Study *The New Zealand Statistician* 18 (2): 20–33.

Renner, R.M. (1984) Rejoinder *The New Zealand Statistician* 19 (1): 31–2.
Roke, C. (18 April 1990) Interview with the author.
Sun, M. (23 November 1984) Panel Says Depo-Provera Not Proved Safe *Science* 226: 950–1.
Unattributed (1991) DMPA and breast cancer: the dog has had its day *Lancet* 338: 856–7.
Upjohn Company (1986) Depo-Provera Issue Discussion Paper.
Weisz, J., Ross, G.T., and Stolley, P.D. (1984) *Report of the FDA Public Board of Inquiry on Depo-Provera* Rockville, Maryland, United States Food and Drug Administration.
World Health Organization (WHO) (1986) Depot-medroxyprogesterone acetate (DMPA) and cancer: Memorandum from a WHO meeting *Bulletin of the WHO* 64 (3): 375–82.
WHO Collaborative Study of Neoplasia and Steroid Contraceptives (1985) Invasive cervical cancer and depot-medroxyprogesterone acetate *Bulletin of the WHO* 63 (3): 505–11.
WHO Collaborative Study of Neoplasia and Steroid Contraceptives (1991) Breast cancer and depot-medroxyprogesterone acetate: a multinational study *Lancet* 338 (8771): 833–8.

The Mass Treatment Trap: Hypertension and Hypercholesterolaemia

Rodney Jackson and Ichiro Kawachi

Introduction

In 1978 the long-awaited results of the World Health Organization (WHO) Clofibrate Trial, a major trial of the prevention of coronary heart disease using the cholesterol-lowering drug clofibrate, were published in the *British Heart Journal* (Committee of Principal Investigators, 1978). In this trial the occurrence of non-fatal myocardial infarction (heart attack) was significantly reduced in the group taking the drug clofibrate. This was the result the investigators had hoped to find. However, unexpectedly and of considerable concern, was the finding that the overall death rate from all causes was also significantly higher in the clofibrate group. The study population consisted of male volunteers with moderately high blood cholesterol, but free of heart disease at the time of entry into the trial. The risk of death from heart disease in this group without treatment was only 2.9 per 1000 per year, and it appears that the small increased risk of death associated with a higher blood cholesterol was outweighed by a slightly larger, although still small, risk of death due to adverse effects of the drug. A small risk of death associated with taking a drug becomes important when it outweighs the benefits of the drug, and this can occur when the risk of death associated with the condition to be treated is also very low. When a drug is recommended for mass treatment, as is now the case for the treatment of high blood cholesterol, many of those identified for treatment will have only a small increased risk of death. In this situation the benefits of treatment can be relatively easily outweighed by a small adverse effect of treatment; this is 'the mass treatment trap'.

This chapter examines the mass treatment trap in relation to hypertension (high blood pressure) and hypercholesterolaemia (high blood cholesterol). Hypertension is already treated on a mass scale in a number of countries, and with the introduction of new, effective cholesterol-lowering agents it appears likely that the management of hypercholesterolaemia will follow a similar path. In developed countries safety regulations for pharmaceuticals aim to minimize the release of drugs with frequent major side-effects. However, the increased risk of death associated with drugs such as clofibrate (less than two per 1000 persons per year) is probably too low to be detected in the pre-marketing studies carried out to meet safety regulations.

To better understand the mass treatment trap it is worth digressing on the notion of 'risks'. There are two major ways of looking at risk — absolute risk and relative risk. *Absolute* risk is the chance of something happening — for example, it is the risk of dying of heart disease in the next 12 months if you are a man aged 45 years and happen to have a high serum cholesterol level of 8.0 mmol/L (that is, eight millimoles of cholesterol per litre of blood). Unfortunately, it is easy for a professional to form a biased view of absolute risk. Take the following example, given by Professor Stephen Leeder (1988):

A woman who had fibroids (a benign tumour of the womb) . . . required a hysterectomy. Her consultant gynaecologist said to her: 'Of course, we'll take the ovaries at the same time'. 'Why?' she asked. He replied: 'I've seen too many cases of carcinoma [cancer] of the ovary. It's very hard to detect, you know. You wouldn't want to die of that, would you?' The woman went ahead and had her ovaries removed.

In fact, at the age of 38 the woman's absolute risk of developing cancer of the ovary was about 14 in 1000, in other words, 986 out of every 1000 women have nothing to worry about. It is just that the gynaecologist sees a lot of cancer of the ovary — this is his professional bias, his distorted perception of risk (Leeder, 1988).

Another way of looking at any particular risk is to compare it with some other risk. This is called relative risk. *Relative* risk compares the risk in one group of individuals exposed to a particular risk factor (for example, raised cholesterol) with the risk in a similar group that is not exposed to that risk factor (for example, a group with 'normal' cholesterol). For example, the relative risk of suffering a non-fatal heart attack for a 40-year-old man with a serum cholesterol of 8 mmol/L is approximately 1.8 compared to another 40-year-old man with 'normal' cholesterol (this being about 5.6 mmol/L in the New Zealand male population of this age). Translated into plain English, the relative risk of 1.8 means that the man with raised cholesterol is 1.8 times as likely to get a heart attack as the man with normal cholesterol.

The Potential for Mass Treatment

Distribution of risk factors

Figure 1 shows the distribution of systolic blood pressure in two very different populations: London civil servants and Kenyan nomads (Rose, 1985). It is noteworthy that despite major differences in blood pressure levels between the two populations, the shape of the distribution curves is similar. There is no obvious cut-off level of blood pressure to distinguish between subjects with 'normal' as opposed to 'elevated' blood pressure. Moreover, if London civil servants with blood pressure levels in the top 10 per cent of the distribution were defined as 'hypertensive', the cut-off level would be above the highest blood pressure of any Kenyan nomad. Conversely, the cut-off blood pressure for the top 10 per cent of Kenyan nomads would be lower than the median blood pressure among London civil servants. It is therefore not easy to define what level of blood pressure is harmful.

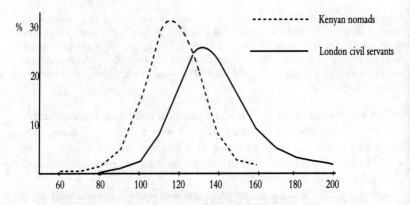

Figure 1: Distributions of systolic blood pressure in middle-aged men in two populations (After Rose, 1985).

Figure 1: Distributions of the systolic blood pressure in middle-aged men in two populations (after Rose, 1985).

The distribution of cholesterol levels between South Japan and East Finland shows exactly the same pattern as for blood pressure (WHO Expert Committee, 1982). Similarly, transferring definitions of hypercholesterolaemia based on cholesterol levels from one population to another has major practical implications. Shaper and Pocock (1985) have shown that applying United States definitions of hypercholesterolaemia ('high' risk defined as serum cholesterol of 6.72 mmol/L) to United Kingdom populations would classify 31 per cent of UK men as hypercholesterolaemic — having high blood cholesterol — whereas the same definition so classifies only 10 per cent of US men into this category.

Small changes in definitions of hypertension or hypercholesterolaemia within a population also have major effects on the proportion of a population classified as eligible for treatment. For example, definitions of hypertension using diastolic blood pressure levels of 100 mmHg (millimetres of mercury), 95 mmHg and 90 mmHg would classify, respectively, about 11 million, 23 million, and 40 million adult Americans as hypertensive (Moore, 1990).

Disease risk

Figure 2 shows the age-adjusted coronary heart disease and total six-year death rate per 1000 men screened for the Multiple Risk Factor Intervention Trial (MRFIT) according to cholesterol and blood pressure classified by percentiles (percentage points) of distribution (Martin et al., 1986). The disease risks associated with elevated cholesterol and blood pressure follow an identical pattern.

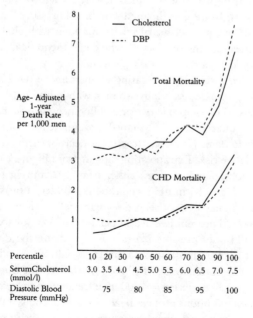

Figure 2

The risk of coronary heart disease death increases linearly, albeit slowly, from the lower end of the distribution until about the 80th to the 85th percentile. In this population the slope of the risk curve for coronary heart disease then becomes progressively steeper.

The pattern is similar for overall mortality, except for a small upturn in risk below the 15th percentile. The major upturn in both total and coronary heart disease mortality around the 85th percentile (that is, at a serum cholesterol level of around 253 mg/dL [6.7 mmol/L] and a diastolic blood pressure level at around 94 mmHg) suggests a natural cut-off point for increased mortality. However, as the risk of both overall mortality and coronary heart disease mortality increases from the 15th percentile (that is, at a serum cholesterol level of around 175 mg/dL [4.6 mmol/L], and a diastolic blood pressure level at around 73 mmHg), there could also be some justification for taking this level as a cut-off point for recommending treatment.

Small changes in the definition of 'hypertension' within a population can have a major influence on the proportion of a population labelled as eligible for treatment. It could be argued, based on the MRFIT data (Figure 2), that up to 85 per cent of American men aged 37–54 years could benefit from drug treatment which lowered their blood pressure or cholesterol levels. Indeed, as the men with blood pressure levels above the 15th percentile are not necessarily the same men with cholesterol levels above this percentile, the proportion who could theoretically benefit from treatment would be *higher* than 85 per cent.

The suggested rationale for treating all men above the 15th percentile is based on their disease risk relative to men with lower cholesterol or blood pressure levels. This approach, using differences in *relative* risk as the justification for intervention, has shaped most guidelines for the treatment of risk factors to date. However, treatment decisions should be based on *absolute* not relative risk. Extrapolating directly from Figure 2 indicates that reducing diastolic blood pressure either from 102 mmHg to 94 mmHg or from 94 mmHg to 84 mmHg would both produce approximately a 30 per cent reduction in coronary heart disease mortality. However, the former intervention would prevent five deaths per 1000 treated for six years, while the latter would only prevent 2.5 deaths. Reducing the diastolic blood pressure level from 84 mmHg to 65 mmHg would also produce a 30 per cent reduction in coronary heart disease mortality, but translates into a saving of only 1.5 lives for every 1000 men treated for six years. Although each of these interventions reduces heart disease mortality by 30 per cent, the benefits of reducing blood pressure levels at the top end of the distribution are clearly the greatest. Moreover, if the treatment increased non-heart disease deaths by two per 1000 per six years, then an overall mortality reduction would only be achieved for the groups with initial blood pressure levels of 102 mmHg and 94 mmHg. The group with initial blood pressure levels of 84 mmHg would experience a small increase in mortality (an example of the mass treatment trap).

Interventions — the Benefits and Risks

Hypertension

A recent pooled analysis of 14 randomized trials of antihypertensive therapy indicated that a reduction in diastolic blood pressure of 5–6 mmHg for approximately five years was associated with a reduction in the risk of stroke of 42 per cent and coronary heart disease risk of 14 per cent (Collins et al., 1990). In the subset of trials in which all patients had diastolic blood pressure levels less than 110 mmHg (that is, with 'mild' hypertension), the reduction in stroke was 41 per cent and in heart disease was 10 per cent.

For the 'mildly' hypertensive group this substantial relative reduction in risk translates into an absolute reduction in the number of fatal and non-fatal stroke events by only six per 1000 patients treated for five years and in the numbers of coronary heart disease events by only 3.3 per 1000 patients treated for five years. Moreover, for every 1000 patients with mild hypertension treated per year, only one death was prevented overall.

In most of the trials reviewed the observed reduction in stroke events was equal to the expected reduction based on the epidemiological data, although only half the expected reduction in coronary heart disease events occurred. Furthermore, if we exclude the Hypertension Detection and Follow-up Program (1979) which was 'unblinded' (i.e. the treatment was not disguised), the effect of lowering blood pressure on heart disease risk in mildly hypertensive patients is minimal. Although it is possible that the studies were too short to show the effect of reducing blood pressure on heart disease risk, it has been postulated that the antihypertensive drugs commonly used in the trials — such as diuretics and beta-blockers — may have had some adverse as well as beneficial effects (Collins et al., 1990).

Thiazide diuretics (one of the commonly-used drugs for reducing blood pressure) are known to produce falls in blood potassium. Low blood potassium has been shown to be significantly related to the risk of heart disease, possibly due to its effect of producing rhythm disturbances of the heart (MacMahon et al., 1986). Diuretics are also known to increase blood lipid levels and have adverse effects on glucose metabolism (Pollare et al., 1989). Beta-blockers, the other commonly-used antihypertensive drug, reduce

HDL cholesterol (the so-called 'good' cholesterol) and increase triglyceride (MacMahon et al., 1986). Newer drugs, for example angiotensin-converting enzyme (ACE) inhibitors and calcium channel blockers, do not appear to have adverse effects on either lipids or glucose metabolism (Pollare et al., 1989). However, there is as yet no clinical trial evidence that the new generation drugs have any beneficial effects on cardiovascular disease (heart and stroke) risk or total mortality. Moreover, there is some evidence that calcium channel blockers, when used after a heart attack, are associated with increased total mortality compared with an inert placebo (Held et al., 1989).

The concerns raised about diuretics and beta-blockers highlight the need for studies using reductions in cardiovascular disease and total mortality as criteria, rather than intermediate measures such as falls in blood pressure. Until randomized controlled studies show that the new generation antihypertensives reduce cardiovascular disease morbidity and mortality (and ideally total mortality), there is little justification for their use, except when diuretics and beta-blockers are clearly contra-indicated. Despite the lack of evidence of real benefit, the new generation antihypertensives have been aggressively marketed and they are now commonly prescribed (Moser et al., 1991).

In addition to their adverse effects, excessive lowering of blood pressure with drugs may paradoxically increase rather than decrease heart disease risk in some groups of individuals. There is a growing body of clinical and experimental evidence to suggest the existence of a relationship between excessive lowering of blood pressure and increased risk of heart attack, particularly among individuals with pre-existing heart disease (Cruickshank, 1988). For example, of the participants in the Hypertension Detection and Follow-up Program with mild hypertension, those who had a diastolic blood pressure reduction of more than 10 mmHg experienced a slightly increased risk of death compared with those who had a smaller decrease in diastolic blood pressure (Cooper et al., 1988).

If any of the above-mentioned adverse effects were independent of the initial heart disease risk of treated individuals then, among low-risk individuals (such as women), the benefits of blood pressure lowering on heart disease risk could be outweighed by the risk associated with the adverse drug effects. Evidence suggests this may be the case. Indeed, in both trials a small increase in total mortality was observed in treated white women who were at the lowest absolute risk.

In addition to the possible adverse effects on disease outcomes, a substantial number of side-effects of drugs were reported in both the MRC (Medical Research Council) Trial (MRC Working Party, 1981) and the

Hypertension Detection and Follow-up Program (Curb et al., 1985). In the MRC trial, treatment with two drugs — bendrofluazide and propranolol — was significantly associated with impotence in men. Other significant side-effects of drugs included gout and diabetes with bendrofluazide, and Raynaud's phenomenon and shortness of breath with propranolol. Approximately 17 per cent of men and 13 per cent of women withdrew from treatment with bendrofluazide due to suspected side-effects during the first five years of the study (MRC Working Party, 1981). The withdrawal rates for propranolol were 15 per cent and 18 per cent for men and women respectively. The Hypertension Detection and Follow-up Program reported similar adverse effects of treatment (Curb et al., 1985).

Withdrawal of patients with suspected adverse drug reactions from the *placebo-treated* groups in many antihypertensive drug trials is also common. This suggests that the act of labelling people as being at risk of disease can itself have adverse effects, aside from specific treatment effects. Bloom and Monterosa (1981) followed a group of individuals mislabelled as hypertensive in a prevalence survey. None of the group were subsequently treated, although they reported significantly more depressive symptoms and poorer health than a matched control group of patients. Other studies also suggest that screening and identification of risk factors can have adverse psychological effects (Stoate, 1989).

Although no deaths were attributed to adverse reactions of treatment in the Hypertension Detection and Follow-up Program or the MRC Trial, the occurrence of side-effects severe enough to result in withdrawal from drug therapy was clearly common. The potential importance of these 'minor' side-effects is illustrated in a paper by Edelson and colleagues (1990) on the cost-effectiveness of hypertension treatment. They estimated that despite only a 2 per cent drop in the patient's quality of life each year through taking antihypertensive medicines, the risks of treatment would outweigh the benefits, implying that the patient would be better off without taking drug treatment.

There is limited evidence that the new generation antihypertensives have fewer side-effects than diuretics and beta-blockers (Croog et al., 1986). Small differences in side-effect profiles of drugs could have an important influence on the balance of risks and benefits of treating mild hypertension. However, because of the price difference between, for example, beta-blockers and ACE inhibitors, beta-blockers remain the most cost-effective treatment even after adjustment for differences in side-effects (Edelson et al., 1990).

In summary, lowering blood pressure with drugs in patients with mild hypertension appears to reverse the risk of stroke within a few years but has a disappointingly small, if any, beneficial effect on coronary heart disease risk. The lack of benefit for heart disease risk is likely to be due in part to adverse effects of treatment. Taken together with the incidence of other side-effects as described above, it is likely that for treated hypertensives at low absolute risk of cardiovascular disease, for example white women with mild hypertension, there will be little or no gain from treatment. This is an example of the mass treatment trap.

High blood cholesterol

In the WHO clofibrate trial a mean reduction in serum cholesterol of approximately nine per cent was achieved in the clofibrate-treated group. This was associated with a 20 per cent reduction in the incidence of coronary heart disease, although the fall was confined to non-fatal heart attack (Committee of Principal Investigators, 1978). A major cause for concern in this study was the significant excess in total deaths in the *treated* compared with the placebo group (162 vs 127). The excess in deaths was mainly confined to diseases related to the liver, the biliary, and the intestinal systems. Therefore, despite a favourable effect on heart disease, the adverse effects of the drug clearly outweighed the benefits in this relatively low risk population with a mean pre-intervention serum cholesterol of only 6.4 mmol/L and a coronary heart disease mortality rate without treatment of only 2.9 per 1,000 persons per year. It is therefore not surprising that a small adverse effect of treatment overwhelmed any benefits of prevention. A subsequent long-term follow-up of the study participants (mean 9.4 years: 5.3 in trial and 4.3 afterwards) showed a statistically significant 25 per cent excess in total deaths in the treated group compared with placebo (Committee of Principal Investigators, 1980). No particular disease accounted for the overall excess.

Muldoon and colleagues (1990) recently conducted a pooled analysis of six heart disease prevention trials of cholesterol lowering which included 24,847 patients and approximately 119,000 person-years of follow-up (mean treatment duration 4.8 years). Their analyses supported previous reviews showing that lowering cholesterol is associated with a reduction in heart disease mortality. However, they also identified a significant increase in deaths unrelated to heart disease such as accidents, suicide or violence, as well as a possible increase in cancer. Overall, cholesterol lowering was *not* associated with a reduction in total mortality.

A more comprehensive pooled analysis of 19 trials of coronary heart disease prevention, representing over 100,000 participants, was undertaken by Holme (1990). With respect to cholesterol lowering and risk of coronary heart disease, the observed reduction in risk was similar to the expected reduction based on epidemiological studies. However, what was gained by cholesterol lowering appeared to be mostly lost as a result of other causes of death. In Holme's review there was *no* overall reduction in total mortality in the treated groups.

Yusuf et al. (1988) described an unpublished review of 22 randomized trials of cholesterol lowering which also shows no reduction in total mortality despite significant reduction in fatal and non-fatal coronary heart disease. With the exception of the clofibrate trial, Yusef et al. suggested that the increase in non-coronary heart disease death was most likely due to chance. In a joint statement from the American Heart Association and the National Heart, Lung and Blood Institute, *The Cholesterol Facts* (Gotto et al., 1990), no specific comment at all was made concerning the possible adverse effects of treatment. It was also correctly pointed out that the major primary prevention trials were not designed to demonstrate a decline in total mortality. It was further noted that in the long-term follow-up of three trials — The Coronary Drug Project (Canner et al., 1986), the Oslo Study Diet and Anti-smoking Trial (Hjermann et al., 1981), and the Stockholm Ischaemic Heart Disease Trial (Carlson and Rosenhamer, 1988) — a reduction in total mortality has been reported.

Interestingly, two of these studies were so-called 'secondary prevention' trials; that is, trials in which the patients already had pre-existing heart disease, and thus were at high absolute risk of further heart attacks (for example, the coronary heart disease mortality rate in the Coronary Drug Project was 36 per 1000 persons per year, and in the Stockholm Trial, 53 per 1000 persons per year). The coronary heart disease mortality rate in the Oslo Trial was considerably lower (4.4 per 1000 persons per year). However, the treatment included both diet and anti-smoking advice in a group with both high serum cholesterol levels (greater than 7.5 mmol/L) and high smoking prevalence (80 per cent). These studies were less susceptible to the mass treatment trap than the three major coronary heart disease prevention trials described below, in which the absolute risk of coronary heart disease in the untreated group was low and the treatment only moderately successful at reducing risk. The untreated participants in the WHO clofibrate trial had an annual heart disease mortality rate of only 1.2 per 1000 persons per year (Committee of Principal Investigators, 1978). The rate was also low in the Lipid Research Clinics Coronary Primary Prevention Trial (1984), at 3 per 1000 persons per year, and in the Helsinki Heart Study, at 1.9 per 1000 persons per year (Frick et al., 1987).

Moreover, the 9 per cent, 8.5 per cent and approximately 10 per cent drop in serum cholesterol levels in the treated compared with control groups of these three trials respectively did not appear to reduce the risk of coronary heart disease mortality enough to counter-balance possible adverse effects of cholesterol lowering. This is consistent with the findings of Holme (1990) who, in his pooled analysis of cholesterol lowering trials, suggested that at least an 8 to 9 per cent cholesterol reduction had to be achieved before an associated reduction in total mortality would occur. Holme, however, did not mention that the pre-treatment absolute risk of coronary heart disease was equally, if not more, important than the actual degree of cholesterol lowering achieved in determining whether total mortality would be reduced.

There is little published data on the less serious side-effects of lipid-lowering treatment. Cholestyramine, one of the mainstays of treatment until recently, is well known to be poorly tolerated by many patients. In the Lipids Research Clinics Trial, 27 per cent of the participants in the cholestyramine-treated group were either unwilling or unable to take an adequate amount of medication in the final year of the study (Lipid Research Clinics Program, 1984). Treatment with niacin, which significantly reduced total mortality in the Coronary Drug Project (Canner et al., 1986), causes numerous minor side-effects. Up to 92 per cent of participants reported uncomfortable rashes (49 per cent reported itching, and 20 per cent reported other rashes) (Moore, 1989). Minor side-effects are much less commonly reported with other cholesterol-lowering drugs such as the fibrates and the recently-introduced co-enzyme A reductase inhibitors.

In summary, treatment appears to largely reverse the risk of coronary heart disease attributed to high cholesterol. However, most trials show non-significant increases in non-heart disease mortality which *counter-balance* any beneficial effects of treatment. The only studies showing significant reductions in total mortality included participants at high absolute risk of coronary heart disease, or involved multiple interventions (such as smoking cessation or dietary advice).

Treatment Guidelines

Despite the disappointing results of trials of blood pressure and cholesterol lowering in individuals at low absolute risk of cardiovascular disease (that is, stroke and heart disease), national and international treatment guidelines take limited account of absolute risk. The 1988 Report of the Joint National Committee on Detection, Evaluation, and Treatment of High Blood Pressure in the United States states that:

The benefits of drug therapy appear to outweigh any known risks to individuals with persistently elevated diastolic blood pressure greater than 94 mmHg (Joint National Committee, 1988).

They add that people with diastolic blood pressure levels between 90 and 94 mmHg who are otherwise at high risk should be treated with drugs if non-drug therapy fails. The report also states that some experts believe all people with diastolic blood pressure levels persistently above 90 mmHg should be treated with drugs. Despite the attention to other risk factors at diastolic blood pressure levels below 95 mmHg, treatment is recommended for all men and women aged 25 to 64 years (and probably for older hypertensive patients as well) if their diastolic blood pressure levels are above 94 mmHg. As indicated in the present chapter, such advice *cannot* be justified on the available evidence.

Moreover, based on Framingham Study data (American Heart Association, 1973), this advice means that two people with a 30-fold difference in coronary heart disease risk over the following six years — for example, a 45-year-old woman whose only risk factor is a raised diastolic blood pressure, and a 65-year-old man with multiple risk factors — could be offered the same treatment. Similarly, the United States National Cholesterol Education Program guidelines advise individualized treatment for all Americans aged 25 to 74 years with low-density lipoprotein (LDL — the so-called 'bad' cholesterol) levels above 160 mg/dL (4.1 mmol/L), regardless of their absolute risk of coronary heart disease (National Cholesterol Education Program Expert Panel, 1988). At LDL cholesterol levels between 130 and 160 mg/dL, other risk factors are taken into account before treatment is advised. Nevertheless, it is difficult to justify similar individual treatment guidelines for a 45-year-old woman with an LDL cholesterol of 165 mg/dL but no other coronary heart disease risk factors (six-year coronary heart disease risk of one in 1000) and a 65-year-old male with the same LDL level but who smokes and has other serious risk factors (six-year coronary heart disease risk of 41.8 in 1000) (American Heart Association, 1973).

It is not surprising that such vast discrepancies exist in the absolute risk of cardiovascular disease between individuals with different risk factors, given that the National Cholesterol Education Program guidelines would identify approximately 36 per cent of all Americans aged 25 to 74 years as eligible for treatment (Sempos et al., 1989). The Joint National Committee (1988) guidelines identify a similar proportion of Americans as requiring antihypertensive treatment. In view of the major differences between individuals in absolute cardiovascular disease risk, these recommendations are clearly inappropriate and require radical revision. Current recommendations have made risk factor reduction an end in itself. Instead, guidelines should recommend that physicians assess the absolute risk of cardiovascular disease in each patient using algorithms (or estimation procedures) such as the Framingham Study equations which incorporate information on age, sex, smoking status, blood pressure, serum cholesterol, and diabetes (Gordon et al., 1971). These are now available on inexpensive hand-held calculators and could easily be used in clinical practice.

Treatment guidelines should apply equally to pharmacological and non-pharmacological interventions, as the latter are not without costs and possible risks. Weight reduction, for example, can be an effective treatment for hypertension; however, the costs to the individual of preparation of special diets, therapy, and foregoing the pleasures of certain foods may outweigh the benefits of weight reduction in those at low risk of hypertension-associated disease. Similarly, strict diets low in saturated fat will generally reduce cholesterol levels, but again there are costs. Moreover, it has been suggested that lowering blood cholesterol by any means may cause a small increase in the risk of death from cancer, accidents, violence, and suicide (Muldoon et al., 1990).

The major challenge for the medical profession, and indeed society, will be to decide what level of absolute risk is worth treating. This may vary depending on the values and resources of the individual patient or the community. The cost-effectiveness of alternative interventions is increasingly regarded as an important aid to this decision-making process, with currently accepted practices initially used as bench-marks. While the cost-effectiveness approach has yet to be widely accepted or incorporated in the process of health policy formulation, it nevertheless represents a significant improvement on the often arbitrary decisions which have been made by national and international committees in the past.

The Mass Treatment Trap and the Pharmaceutical Industry

The examples of hypertension and elevated blood cholesterol suggest that there are certain pre-conditions which need to be met before the mass treatment trap occurs. The availability of effective treatment is, needless to say, a *sine qua non*. However, this raises the question of how the pharmaceutical industry initially made the decision to invest in research to develop these drugs. Hypertension and high blood cholesterol in fact share certain characteristics which shed light on the genesis of the mass treatment trap.

The most important pre-condition of the mass treatment trap is that the treatable 'disease' must affect a large proportion of the general population. The greater the proportion of the population affected, the greater the potential market for a remedy, and hence the more likely that the pharmaceutical industry will invest in its research. As noted earlier, hypertension and high serum cholesterol are both 'diseases' which affect potentially vast numbers in the general population, depending on the way that 'disease' is defined.

The second pre-condition for the mass treatment trap is that there should be no discernible *threshold* of risk. As explained earlier, in the case of both hypertension and high blood cholesterol, the risk of heart disease and stroke rises continuously and in a curvilinear manner from the lowest levels (including people with 'normal' blood pressure and blood cholesterol) to the highest. This lack of a natural threshold in risk means that there are opportunities for continuously redefining the level at which there is a 'need' for drug treatment. Each time the committees of experts convene to set a new — and lower — threshold of treatment, the size of the drug market expands. This is exactly what has happened in the case of hypertension, where, beginning in the 1960s and throughout the 1970s, the cut-off levels of blood pressure 'in need of treatment' were successively lowered by various expert committees. The result has been a massive, but predictable, rise in the number of people on antihypertensive medication. Elevated serum cholesterol appears set to follow a similar pattern. Indeed, a shrewd market analyst would advise the pharmaceutical industry to invest money into exactly those areas where there is opportunity for future market expansion.

The third pre-condition for mass treatment is that the disease should ideally be 'incurable'; that is, drug treatment should be life-long, thereby ensuring the greatest possible volume of sales. For example, hypertension is said to be 'incurable' in the sense that without regular medication the blood pressure often creeps back up. Drug treatment is therefore most often recommended for an indefinite period. Interestingly, treatment for such 'incurable' conditions also tends to thrive in health care systems based upon a fee-for-service form of payment. This may partly explain, for example, the fact that physicians in the United States adopt a much more aggressive approach to the treatment of hypertension and elevated blood cholesterol compared to their British counterparts (who are mostly salaried workers). There is evidence to suggest that United States physicians generally treat both these conditions at lower thresholds of risk than in the United Kingdom (Weiland et al., 1991).

The fourth pre-condition for mass treatment is that opportunities should exist for spin-off industries. In the case of elevated serum cholesterol, there are spin-off benefits for the screening industry (including private 'laboratories' set up in shopping malls), as well as the food industry (marketing low-fat, no-cholesterol products). When diverse industrial interests coalesce in this way, it becomes much more difficult to stem the tide of mass treatment.

Perhaps the most persuasive argument in defence of mass treatment is that it constitutes a strategy of prevention. By masquerading as a preventive activity, mass treatment is able to stay congruent with the ideology that says 'prevention is better than cure'. However, as discussed earlier in the chapter, prevention is not necessarily better than cure, especially among people at low absolute risk of disease. Enough is now known about the potential harm of mass treatment to warrant a more cautious approach to risk factor modification in the general population.

Conclusion

An intervention can only be justified if the benefits are believed to outweigh the risks and costs of the intervention. The view that a treatment can be justified just because a benefit — no matter how small — has been demonstrated, is becoming less acceptable (Rose, 1990). All interventions have associated risks and costs and these become particularly important when the potential absolute benefits are small and can be outweighed by a small risk. The current fashion of mass screening and treatment of individuals with risk factors for cardiovascular disease in a number of developed countries, particularly in the United States, potentially exposes large numbers of people with a low absolute risk of disease to treatments which, at best produce little or no benefit, and at worst cause more harm than good. This is the mass treatment trap. The interventions most susceptible to the mass treatment trap are those which involve on-going treatment and therefore on-going risks and costs such as the individualized treatment of hypertension and hypercholesterolaemia described in this chapter. The mass treatment trap can be avoided if physicians explicitly assess the absolute risk of disease and the absolute potential for benefit before embarking on a mass screening or treatment programme. This will require radical revision of national and international treatment guidelines. Guidelines for *individual* risk factors are inappropriate in the context of mass screening and treatment, and need to be replaced by one set of guidelines for cardiovascular disease control which incorporate *all* risk factors into one decision-making process.

References

American Heart Association (1973) *Coronary Risk Handbook* American Heart Association.

Bloom, J.R. and Monterossa, S. (1981) Hypertension labelling and sense of well-being *American Journal of Public Health* 71: 1228–32.

Canner, P.L., Berge, K.G., Wenger, N.K. et al. (1986) Fifteen year mortality in Coronary Drug Project patients: long-term benefit with niacin *Journal of American Coronary Care* 8: 1245–55.

Carlson, L.A. and Rosenhamer, G. (1988) Reduction of mortality in the Stockholm Ischaemic Heart Disease Secondary Prevention Study by combined treatment with clofibrate and nicotinic acid *Acta Medica Scandinavica* 233: 405–18.

Collins, R., Peto, R., MacMahon, S. et al. (1990) Blood pressure, stroke, and coronary heart disease. Part 2, short-term reductions in blood pressure; overview of randomised drug trials in their epidemiological context *Lancet* 335: 827–38.

Committee of Principal Investigators (1978) A co-operative trial in the primary prevention of ischaemic heart disease using clofibrate *British Heart Journal* 40: 1069–118.

Committee of Principal Investigators (1980) WHO Co-operative trial on primary prevention of ischaemic heart disease using clofibrate to lower serum cholesterol: mortality follow-up *Lancet* ii: 379–85.

Cooper, S.P., Hardy, R.J., Labarthe, D.R. et al. (1988) The relation between degree of blood pressure reduction and mortality among hypertensives in the Hypertension Detection and Follow-up Program *American Journal of Epidemiology* 127: 387–403.

Croog, S.H, Levine, S., Testa, M.A. et al. (1986) The effects of antihypertensive therapy on quality of life *New England Journal of Medicine* 314: 1657–64.

Cruickshank, J.M. (1988) Coronary flow reserve and the J-curve relation between diastolic blood pressure and myocardial infarction *British Medical Journal* 297: 1227–30.

Curb, J.D., Borhani, N.O., Blaszkowski, T.P. et al. (1985) Long-term surveillance for adverse effects of antihypertensive drugs *Journal of the American Medical Association* 253: 3263–8.

Edelson, J.T., Weinstein, M.C., Tosteson, A.N.A. et al. (1990) Long-term cost-effectiveness of various initial monotherapies for mild to moderate hypertension *Journal of the American Medical Association* 263: 408–13.

Frick, M.H., Olli, E., Haapa, K. et al. (1987) Helsinki Heart Study: primary-prevention trial with gemfibrozil in middle-aged men with dyslipidaemia *New England Journal of Medicine* 317: 1237–45.

Gordon, T., Sorlie, P., and Kannel, W.B. (1971) Coronary heart disease, atherosclerotic brain infarction, intermittent claudication — a multivariate analysis of some factors related to their incidence: Framingham study, 16-year follow-up. Section 27, US Government Printing Office.

Gotto, A.M., LaRosa, J.C., Hunninghake, D. et al. (1990) The cholesterol facts *Circulation* 81: 1721–33.

Held, P.H., Yusef, S., and Furberg, C.D. (1989) Calcium channel blockers in acute myocardial infarction and unstable angina: an overview *British Medical Journal* 299: 1187–92.

Hjermann, I., Velve Byre, K., Holme, I., and Leven, P. (1981) Effect of diet and smoking intervention on the incidence of coronary heart disease *Lancet* ii: 1303–10.

Holme, I. (1990) An analysis of randomized trials evaluating the effect of cholesterol reduction on total mortality and coronary heart disease incidence *Circulation* 82: 1916–24.

Hypertension Detection and Follow-up Program Co-operative Group (1979) Five-year findings of the Hypertension Detection and Follow-up program *Journal of the American Medical Association* 242: 2562-71.

Joint National Committee (1988) The 1988 Report of the Joint National Committee on Detection, Evaluation and Treatment of High Blood Pressure *Archives of Internal Medicine* 148: 1023-38.

Leeder, S.R. (1988) The estimation and modification of risk in general practice. Part one *Australian Family Physician* 17: 658-61.

Lipid Research Clinics Program (1984) The Lipid Research Clinics Coronary Primary Prevention Trial Results *Journal of the American Medical Association* 251: 351-64.

MacMahon, S.W., Cutler, J.A., Furberg, C.D., and Payne G.H. (1986) The effects of drug treatment for hypertension on morbidity and mortality from Cardiovascular Disease *Cardiovascular Disease* 24 (Supplement 1): 99-118.

Martin, M.J., Hulley, S.B., Browner, W.S. et al. (1986) Serum cholesterol, blood pressure and mortality: implications from a cohort of 361,662 men *Lancet* ii: 933-6.

Medical Research Council (MRC) Working Party (1981) Adverse reactions to bendrofluazide and propranolol for the treatment of mild hypertension *Lancet* ii: 539-43.

Medical Research Council (MRC) Working Party (1985) MRC trial of treatment of mild hypertension: principal results *British Medical Journal* 291: 97-104.

Moore, T. (September 1989) The cholesterol myth *The Atlantic Monthly* 37-70.

Moore, T. (August 1990) Overkill *The Washingtonian* 64-7, 194-204.

Moser, M., Blaufox, M.D., Freis, E. et al. (1991) Who really determines your patient's prescription? *Journal of the American Medical Association* 265: 498-500.

Muldoon, M.F., Manuck, S.B., and Matthews, K.A. (1990) Lowering cholesterol concentrations and mortality: a quantitative review of primary prevention trials *British Medical Journal* 301: 309-14.

National Cholesterol Education Program Expert Panel on Detection, Evaluation and Treatment of High Blood Cholesterol in Adults (1988) *Archives of Internal Medicine* 148: 36-69.

Pollare, T., Littel, H., and Erne, C.A. (1989) A comparison of the effects of hydrochlorothiazide and captopril on glucose and lipid metabolism in patients with hypertension *New England Journal of Medicine* 321: 868-73.

Rose, G. (1985) Sick individuals and sick populations *International Journal of Epidemiology* 14: 32-8.

Rose, G. (1990) Reflections on the changing times *British Medical Journal* 301: 683-7.

Sempos, C., Fulwood, R., Haines, C. et al. (1989) The prevalence of high blood cholesterol levels among adults in the United States *Journal of the American Medical Association* 262: 45-52.

Shaper, A.G. and Pocock, S.J. (1985) British blood cholesterol values and the American consensus (letter) *British Medical Journal* 291: 80-1.

Stoate, H.G. (1989) Can health screening damage your health? *Journal of the Royal College of General Practitioners* 39: 193-5.

Weiland, S.K., Keil, U., Spelsborg, A., Hense, H.W., Hartel, U., Gefeller, O., and Dieckmann, W. (1991) Diagnosis and management of hypertension by physicians in the Federal Republic of Germany *Journal of Hypertension* 9: 131-4.

World Health Organization (WHO) Expert Committee (1982) Prevention of coronary heart disease *Technical Report Series* World Health Organization, Geneva.

Yusuf, S., Wittes, J., and Friedman, C. (1988) Overview of results of randomized clinical trials in heart disease *Journal of the American Medical Association* 260: 2259-63.

The Pressure to Treat — Doctors, the Pharmaceutical Industry and the Mass Treatment of Hypertension

Ichiro Kawachi

The Treatment of Hypertension: Past and Present

The development of effective drugs to lower blood pressure is justly regarded as one of the triumphs of modern medicine. The prognosis in the more severe grades of hypertension before the advent of pills in the 1950s was truly appalling. 'Malignant' hypertension — seldom seen these days — was characterized by a rapid onset of heart failure, pulmonary oedema, and stroke or kidney shutdown. Reminiscing about a bygone era in the medical treatment of hypertension, Sir Colin Dollery wrote:

Some patients with the most refractory hypertension had bilateral adrenalectomy (surgical removal of the adrenal glands) and replacement with cortisone. The aim was to find a precarious ledge between death and addisonian crisis on the one hand and malignant hypertension on the other (Dollery, 1987).

Such desperate measures contrast with the cornucopia of blood pressure-lowering drugs available on the market today — over 100 brands of antihypertensive drugs are currently sold in New Zealand alone.

Not only has the treatment changed, but the patients with 'hypertension' have also changed during the forty years since effective drugs first became available. Indeed, the contrast between the clinical picture of 'hypertension' prior to the advent of effective drugs, and 'hypertension' as encountered in modern medical practice, is so dramatic as to make it seem scarcely believable that the doctor is dealing with the same problem. Back in the 1950s and 1960s almost all hypertensive patients were treated for 'malignant' or 'severe' hypertension. Many presented with overt symptoms such as headache, nose bleeds, and breathlessness. They were treated on an urgent basis by specialist physicians in the hospital setting. Today, the vast majority of patients with 'hypertension' are unaware that they have a health problem at the time of diagnosis, and most are detected by the deliberate screening of blood pressure by family physicians.

About 180,000 people in New Zealand are now estimated to be on regular drug treatment for hypertension (Kawachi and Malcolm, 1989). The mass control of hypertension in the general population, made possible by modern medicines, has been hailed as a public health success (Stamler, 1978). However, the mass treatment of the population has not been a free ride. In 1989, blood pressure-lowering drugs and associated medical care costs amounted to $120 million in this country (Malcolm, 1990). Medication has also brought side-effects into many patients' lives, as vividly described by one individual who was diagnosed as having hypertension:

Some six months later methyldopa made its appearance and I was given a very large bottle . . . The heavy dosage was not without its side-effects. Sphincter control became a problem. There was a frequency of bowel motions and an urgency of urination which affected my social life and work. As a doctor I doubt whether I could have asked a patient to endure such treatment . . . [After six months] the methyldopa had been reinforced by chlorothiazide, and for many months I felt weakness and incompetence of thigh and shoulder muscles, which soon disappeared with an aperitif of orange juice. Then came an attack of gout . . . So chlorothiazide was discarded and the newest diuretic substituted. Chlorthalidone was the favourite. The gout disappeared [however] it would be untrue to say that I felt particularly well . . . Over the next twelve months my weight fell gradually from fourteen to eleven and a half stone . . . My nights were disturbed with polyuria. My loss of weight eventually took me to my physician. He found that I had ketonuria . . . I eventually went to hospital, stopped all pills, had my insulin standardised (Anonymous, 1972).

In order to understand how the transition from the 'old' hypertension to the 'modern' type of hypertension came about, it is necessary to trace the history of the development of antihypertensive drugs. The earliest blood pressure-lowering drugs were tested on patients in the 1950s (Restall and Smirk, 1950). Further trials conducted over the two ensuing decades confirmed the efficacy of treating patients with less severe degrees of hypertension (Hennekens, 1987). The Veterans Administration trial was the first to suggest an apparent benefit of drug treatment even among people with 'mild' hypertension (Veterans Administration Cooperative Study Group, 1970). This 1970 report was hailed as 'sounding the death knell of the "therapeutic nihilism" and "judicious neglect" long rampant in the medical profession in regard to this disease'. Calls were made to expunge from the medical lexicon such terms as 'mild' and 'benign' to describe hypertension: 'They are pernicious sources of confusion and misconception — medical anachronisms best relegated to the museum of medical history' (Stamler, 1978). The age of mass treatment of hypertension had begun.

By the mid-1970s the nature of 'hypertension' had completely changed. Many patients judged eligible for treatment now had diastolic pressures of only 100 mmHg or lower and few had any signs or symptoms related to hypertension. The goal of treatment had shifted from the curative (i.e. the treatment of symptomatic individuals) to the preventive (i.e. the forestalling of strokes and heart attacks among asymptomatic populations). The distinction is not merely semantic, though some have decried the inaccuracy of continuing to use the term 'treatment' to describe the management of a risk factor (which is what 'hypertension' has effectively become) (Oliver, 1982). The tangible result of this shift in therapeutic goal from the offensive to the defensive has been observed in terms of the increase in millions of individuals newly diagnosed with 'hypertension' and placed on drug treatment. Hypertension is now the leading indication for visits to the doctor and for the use of prescription drugs (Kaplan, 1990). In the United States, antihypertensive medication was being taken by over 30 per cent of people aged 55 to 64 years, and by over 40 per cent of those aged 65 to 74 years in 1982 to 1984 (Havlik et al., 1989). In New Zealand, where a more conservative attitude towards mass treatment has prevailed, a quarter of the population aged over 65 years was nonetheless estimated to be on regular drug treatment in 1987 (Kawachi and Malcolm, 1989).

Enter the Pharmaceutical Manufacturer

It is perhaps not surprising, given the course of evolution of antihypertensive treatment, that the medical profession has been accused from time to time of 'over-medicalizing' the problem of hypertension (Guttmacher et al., 1981; Berglund, 1984). The sheer number of individuals in the population with 'hypertension' has created a large potential market for blood pressure-lowering drugs. The potential size of the market has not escaped the notice of the pharmaceutical manufacturers, who devote considerable sums of money towards developing a steadily growing armamentarium of 'me-too' drugs to introduce to the market. Moreover, since the advent of safer drugs, the management of hypertension has gradually devolved from specialist physicians to general practitioners, and out of the hospital clinics into the community. This diffusion of antihypertensive therapy from a handful of specialist physicians to thousands of community-based general practitioners significantly enlarged the size of the target audience for the advertising and promotion of various blood pressure-lowering products. Even a cursory examination of medical journals, both 'independent' and drug company-sponsored, reveals the substantial amounts of investment which are being made to advertise antihypertensive drugs to the medical profession.

Apart from the ubiquitous advertisements appearing in medical journals, pharmaceutical manufacturers have employed a variety of other tactics to promote their antihypertensive products. These include: 'reminder' advertisements mailed directly to medical practitioners; the production and circulation (gratis) of journals exclusively devoted to aspects of hypertension management; sponsoring and arranging international meetings or symposia based around their products, which are subsequently serialized as 'supplements' to otherwise 'respectable' and peer-reviewed journals; financing medical researchers to conduct clinical trials on their products, which they later quote as authoritative references to their advertisements; the training and recruitment of medical 'Reps.' who visit doctors and pharmacists on a face-to-face basis; running competitions for doctors and pharmacists, with quizzes based on various claims made about the virtues of the products (the prizes ranging from office 'hardware' to holiday destinations); and the 'gifting' of small items, such as pens, clocks or diaries, emblazoned with the company logo and/or the product brand name (Kawachi and Wilson, 1990).

A recent tactic employed by one pharmaceutical manufacturer in New Zealand was to produce a video promoting their product — an angiotensin-converting enzyme (ACE) inhibitor. The video featured a studio interview with two local experts — a university professor of general practice and a clinical pharmacologist. The pharmaceutical company bought a slot on a national television network, and sent advance notice of the telecast to all practitioners in the country. The advance notice, which resembled a newspaper, proclaimed: 'Tension mounts. Doctors anxiously await vital screening'. Sandwiched between the blank pages was a reply-paid postcard on which each doctor was invited to indicate the preferred hour for a complimentary wake-up call from the company representative on the morning of the telecast. In case any doctor missed the vital screening, a complimentary video-recording of the programme was subsequently sent to every doctor in the country.

Of course such descriptions of promotional gimmicks are not intended in any way to detract from the fundamental achievement of pharmaceutical manufacturers in developing effective drugs to lower blood pressure. As one observer noted:

It is fashionable these days to decry medicinal drugs, and, to some extent, the operations of firms that make them. Certainly I would not wish to defend all the marketing manoeuvres that take place. However, there is no doubt firstly that the drugs that have been developed for the treatment of hypertension have been of immense benefit . . . and secondly that we would not have had any significant drug development if it were not for competition and the profit motive (Simpson, 1990).

Nonetheless, the point at issue is not the effectiveness of blood pressure-lowering drugs *per se*, but the question of whether the current approach to the management of hypertension is the optimal one for the population of New Zealand (or for that matter anywhere else in the world). The thesis of this essay is that the interaction between the medical profession and the pharmaceutical industry has produced a solution to the problem of hypertension which is unsustainable — either philosophically or in practical terms — in the long run.

The Hypertension Controversy

Since the 1970s epidemiologists in several countries have observed a decline in population mortality from stroke and coronary heart disease — two of the most frequent complications of hypertension (Beaglehole et al., 1989; Thom, 1989). In New Zealand, there has been a 50 per cent decline in stroke mortality among the population aged 35 to 74 years over the 30-year period 1954 to 1984 (Bonita and Beaglehole, 1982); and a 20 per cent decline in coronary mortality in the 14-year period since 1968 (Beaglehole et al., 1989). Advocates of hypertension treatment have been quick to associate the observed decline in cardiovascular disease with the widespread detection, treatment and control of hypertension in communities (Levy and Moskowitz, 1982).

Why then, has the last remaining shred of doubt about the value of mass treatment of hypertension yet to be relegated to 'the museums of medical history'? Contrary to expectation, according to one published review of the world literature, the level of debate about the value of such treatment has actually intensified, not diminished, as a result of the accumulation of clinical trial evidence regarding antihypertensive therapy:

The treatment of milder degrees of hypertension has been one of the most controversial issues in medicine during recent years. A virtual cascade of editorials and reviews in both general and sub-specialty journals have debated the interpretation of the large clinical trials of treatment for mild hypertension, with some arguing that they prove aggressive therapy to be effective and virtually mandatory and others arguing that the same studies demonstrate that therapy is too often burdensome and not so clearly effective (Forrow et al., 1988).

The search for an answer to this vexing paradox must start at the final report of the Medical Research Council (MRC) trial of mild hypertension (MRC Working Party, 1985), regarded by many as the definitive study of its kind (it is certainly the largest of the nine trials in existence). The results of this trial, which took nearly a decade to complete from its pilot phase to its final publication, had been eagerly awaited by medical practitioners, policy makers, and pharmaceutical manufacturers. Although numerous guidelines for the treatment of mild hypertension had been issued by various experts prior to the MRC trial, they almost invariably concluded with the temporizing statement that the definitive proof of benefit awaited the publication of the results of that trial (Breckenridge, 1985).

Consequently, the trial results caught everyone by surprise when it was reported that:

If 850 mildly hypertensive subjects are given active antihypertensive drugs for one year about one stroke will be prevented. This is an important but an infrequent benefit. Its achievement subjected a substantial percentage of the patients to chronic side effects, mostly but not all minor. Treatment did not appear to save lives or substantially alter the overall risk of coronary heart disease. More than 95 per cent of the control patients remained free of any cardiovascular event during the trial (MRC Working Party, 1985).

One of the chief investigators of this trial described the press conference in 1985 at which he and his colleagues tried to explain 'to a distinctly unimpressed group of journalists [about] how they had screened half a million people, treated 180,000 of them with very mild hypertension for five years, and had prevented 60 strokes and 5 myocardial infarcts and saved 12 lives' (Dollery, 1987).

The conclusions of the MRC trial appeared to call for a radical reassessment of previous recommendations made by such authoritative bodies as the United States Joint National Committee on Detection, Evaluation and Treatment of High Blood Pressure. For example, in its 1984 consensus statement the Joint National Committee stated that the benefits of drug therapy 'appeared to outweigh any known risks from such therapy for those with diastolic blood pressures persistently elevated above 95 mmHg and for those with lesser elevations who are at high risk' (Joint National Committee, 1984).

Several guidelines issued since the MRC trial have indeed advocated a more conservative approach to the management of hypertension (Sleight, 1985; Hampton, 1986; Memorandum from the World Health Organization/ International Society for Hypertension, 1986; Wilcox et al., 1986; Beaglehole et al., 1988; British Hypertension Society Working Party, 1989). Unfortunately, in spite of such revisions, formidable obstacles still exist which are likely to ensure that the status quo is maintained in the pattern of hypertension management and the delivery of antihypertensive care. Prevailing methods of financing medical care provide incentives for the continued use of medical technologies, regardless of their marginal value. Indeed, as has been noted:

Existing value and belief systems, reward and incentive systems, and patterns of economic and political domination all support the present system. The societal emphasis on curative, technological medicine that has developed over the past decades has its own inertia . . . It is expressed in the medical empires that have grown up around medical schools, with . . . their technologically oriented education for medical students and other future health care providers . . . It is expressed in profits for a large industry and hidden profits of major medical centres and providers. It is expressed in consumer demands for the latest in medical technology. And it is expressed in well-funded political lobbies and political action groups that fund political campaigns (Banta et al., 1981).

In order to understand the nature of the obstacles which stand in the way of resolving the 'dilemma' of hypertension, it is necessary at this point to re-examine the evolution of mass treatment in terms of the structural and institutional interests that have driven it and sustained it.

The Evolution of Mass Treatment

The mass treatment of hypertension presents a fairly typical case study of the diffusion of a medical innovation. McKinlay (1981) described seven stages in the diffusion of a medical innovation. The first stage is the so-called 'stage of the promising report', in which the initial, often spectacular success of an innovation is announced to the public. In the history of antihypertensive medication, the promising reports first appeared in the 1950s. At the time, the effectiveness of the pills at reversing the fatal progression of 'malignant' and 'severe' hypertension was so convincing that there appeared to be no need for the more stringent standards of practice demanded of most clinical trials today, such as the use of an untreated or placebo-treated control group (Hennekens, 1987).

The promising reports were soon followed by the 'stage of professional and organisational adoption' (1960s to early 1970s), during which further clinical trials appeared to confirm the efficacy of antihypertensive treatment among individuals with less severe degrees of hypertension. This period witnessed the launching of major campaigns such as the National High Blood Pressure Education Program (NHBEP) of the National Heart, Lung, and Blood Pressure Institute, an institution which has continued to remain the authoritative policy-maker regarding the treatment of hypertension in the United States and throughout the rest of the world (Levinson and Carleton, 1987). It was during this period also that the threshold level of blood pressure deemed eligible for treatment was successively reset at lower levels by the recommendations of the NHBEP's Joint National Committee (1984). The process of professional adoption continued to be evident, for example in the endorsement of the 1984 consensus statement of the Joint National Committee by 30 national organizations in the United States alone, including the American College of Cardiology, the American College of Physicians, the American Heart Association, and the American Medical Association (Levinson and Carleton, 1987).

Partly as a result of campaigns such as the NHBEP, millions of hypertensive people throughout the world were detected, treated and brought under control — a remarkable achievement considering the traditional attitudes of 'judicious neglect' and 'therapeutic nihilism' that were supposed to have prevailed among the medical fraternity in regard to the treatment of hypertension up until that time (Stamler, 1978). A sociological explanation for this turn-around in medical attitudes has been offered: that in a 'medicalized' society, clinicians frequently feel uncomfortable in merely observing persons identified at risk, and hence treatment may be enthusiastically endorsed despite limited evidence of improved outcome (Brett, 1984). In other words, the availability (and perhaps marketing) of antihypertensive medication created the demand for mass treatment of hypertension. The reports of declining cardiovascular disease mortality which began to appear at about this time may have helped to reassure even the sceptics. Recent analyses, however, cast considerable doubt on the existence of a temporal relationship between the diffusion of antihypertensive therapy and the decline in cardiovascular mortality (Goldman and Cook, 1984; Beaglehole, 1986; Bonita and Beaglehole, 1986; Klag et al., 1989). Mass treatment of hypertension appears to have had a relatively minor impact on the trends in stroke and coronary mortality.

The next stage in the 'career' of a medical innovation — the so-called 'stage of public acceptance and third-party (State) endorsement' — is characteristically associated with an abrupt surge in the diffusion and uptake of the innovation. According to McKinlay, 'An innovation can be said to have "made it" when it eventually receives endorsement or support from the state and/or is underwritten by third parties' (McKinlay, 1981). In New Zealand, where the costs of pharmaceuticals were (until very recently) virtually entirely underwritten by the State, the level of national expenditure on antihypertensive drugs has risen at an alarming rate since the early 1980s. The rising expenditure cannot be explained by an increase in the detection and treatment of hypertensive individuals; in fact, most of the rise in inflation-adjusted expenditure is due to a combination of 'entry effects' (i.e., the listing of new antihypertensive agents on the Drug Tariff) and 'mix effects' (i.e., a change in the pattern of prescribing in favour of more costly drugs) (Kawachi and Malcolm, 1989).

When the patterns of antihypertensive treatment in New Zealand are compared to other countries, some startling differences are revealed. For example, in New Zealand two of the newest and most expensive classes of antihypertensives — the ACE-inhibitors and calcium channel blockers — respectively took up 18.7 per cent and 9.7 per cent of the total market share by volume in 1986 (Kawachi et al., 1989). By contrast, in the United Kingdom and in several European countries, ACE-inhibitors were estimated to take up no more than two per cent of the total market share. The difference is due to the fact that in New Zealand ACE-inhibitors are available for prescription without restriction, whereas in European countries their use is restricted to the most severe and resistant cases of hypertension (Kawachi and Malcolm, 1989). Similarly, calcium channel blockers are freely available for prescription in New Zealand, whereas in the United States their use as an antihypertensive has yet to be approved by the Food and Drug Administration (Gross et al., 1989). Overall, there is little evidence to suggest that hypertensive patients in New Zealand receive better quality care than in other countries; yet differences in third-party endorsement make a profound impact on the diffusion of a product. Unfortunately, 'the state and other third parties do not act on the reasonable basis of reliable evidence, but on the basis of some combination of professional, organisational and public pressure' (McKinlay, 1981). The market shares of these two classes of antihypertensive drugs continue to soar as their manufacturers in New Zealand vigorously promote their use (for example, the ACE-inhibitor lisinopril-MSD was at one time advertised with the caption 'First-line antihypertensive action').

The zenith in the 'career' of antihypertensive therapy was probably reached in the mid-to-late 1970s, when mass treatment of hypertension attained the status of a 'standard procedure'. By then, antihypertensive treatment had become so firmly entrenched throughout the world that in 1978, a survey of New York physicians indicated that 92 per cent of the respondents 'routinely' initiated antihypertensive drug therapy at diastolic blood pressure levels between 90–104 mmHg (i.e., levels of 'mild' hypertension) (Thomson et al., 1981). Since the major trials of treating mild hypertension, such as the Australian trial (Australian National Blood Pressure Study Management Committee, 1980), the MRC trial (MRC Working Party, 1985), and the European Working Party on High Blood Pressure in the Elderly trial (Amery et al., 1985) had not even been completed, common medical practice at the time represented an approach that was even more 'aggressive' than warranted by the available evidence. Thus the fifth stage in the evolution of treating mild hypertension — namely the 'stage of the randomized controlled trial' — followed, rather than preceded its acceptance as a standard medical procedure. By the time the MRC trial results were reported, the mass treatment of hypertension was routine to the extent that between 20-30 per cent of the adult population in several countries including New Zealand were already defined as being 'in need of treatment'; in fact, half of them were actually receiving medication (Jackson et al., 1983).

Given the level of commitment and investment already devoted to the mass treatment of hypertension by the medical profession and the pharmaceutical industry, it is perhaps not surprising that the stage of randomized trials should be followed by 'the stage of professional denunciation' (McKinlay, 1981). As McKinlay observed: 'The success of an innovation has little to do with its intrinsic worth, but is dependent upon the power of the interests that sponsor or maintain it' (McKinlay, 1981). The credibility of the medical profession and the profits of the drug industry appeared to be at stake in the immediate wake of the MRC and other trials. Professional denunciation has since been expressed in numerous letters, articles, and editorials, which have collectively criticized the ostensible shortcomings of the design and analysis of the various trials (Forrow et al., 1988). They have been unanimous in defending the continuation of the mass treatment of hypertension; at the same time ignoring the fact that the results of earlier trials (which were similar to later trials in many respects except that their conclusions were more suited to the detractors' arguments) were once cited as justification for advocating mass treatment in the first place.

On the other hand, doubts about the value of mass treatment have also been vociferously raised. A major concern raised by this camp is the overall balance of the benefits and risks of mass treatment. This is neatly encapsulated in the 'dilemma of hypertension':

The real advantage of treatment for the relatively few who would benefit [is] bought at the expense of the much larger number who would experience the disadvantages of prolonged drug taking and of being labelled as hypertensive, without deriving benefit (Miall, 1986).

Note that the effectiveness of treating hypertension is not being called into question. The point at issue is the 'small but infrequent' benefit, and whether mass treatment by drugs is the most optimal strategy to attain this benefit. As the situation stands at present, little progress is likely to be made towards resolving this dilemma; the dispute about the value of mass treatment of hypertension seems no longer to be a dispute about scientific facts but about values engendered by professional and commercial interests.

For these reasons, the final stage in the 'career' of this medical innovation — 'the stage of erosion and discreditation' —is unlikely ever to be reached. Too many vested interests are involved. Moreover, the fundamental question about the value of mass treatment has been diverted to some extent by the raising of new issues — of 'therapeutic compliance', and 'quality of life'. The pharmaceutical industry has played a major role in these developments.

The Pharmaceutical Industry Again

The treatment of hypertension in the 1980s has been characterized by massive increases in the proportion of the population treated with drugs; spiralling costs of antihypertensive medicines, by the proliferation of 'metoo' formulations entering the market, and by increasing dissent over the optimal strategy for blood pressure control (Kawachi and Wilson, 1990). Manufacturers of antihypertensive drugs have all played their part in these developments. A notable contribution made by the industry in recent years has been their investment of effort and money to elevate 'quality of life' issues in drug therapy to a level of prominence in the debate about mass treatment.

An editorial in 1982 warned that 'insufficient attention [had] been expressed about the incidence of unexpected adverse reactions to drugs given prophylactically over many years in attempts to reduce the risks of coronary and cerebrovascular diseases in predisposed but otherwise healthy persons' (Oliver, 1982). The pharmaceutical industry has apparently taken up this challenge, for they have been actively engaged in promoting the importance of 'quality of life' in antihypertensive therapy, for example, producing journals with such titles as *Quality of Life and Cardiovascular Care* (Wenger, 1984); funding a clinical trial which proved that captopril, an ACE-inhibitor, 'improved' the well-being of patients (Croog et al., 1986); and advertising their products with slogans such as 'Capoten [captopril-Squibb] controls hypertension and gives back your patient's quality of life' and 'There's never been a better time to have hypertension'.

The cumulative effect of such promotion has amounted to a distraction of attention away from the more basic question about whether drug treatment is indicated in the first place. In other words, the industry's apparent concern with 'quality of life' never questions the actual need for drug therapy: 'So long as our pills offer a better quality of life (less impotence, less tiredness) than the rival product' — the argument seems to run — 'then it must be worth swallowing'. This would appear to be the premise of many industry-sponsored clinical trials, including the one which reported that captopril 'improved' patient well-being (Croog et al., 1986). In this trial, patients were randomized to one of three antihypertensive regimens: captopril, propranolol or methyldopa (a drug used widely in the 1960s but seldom prescribed today due to its side-effects). A battery of psychometric measurements were taken on all the participants at baseline, as well as at the end of the trial six months later. Patients receiving captopril apparently 'improved' on their general well-being scores, whereas the others did not. However, closer reading of the Methods section reveals that 75 per cent of all the participants were already on antihypertensive drug treatment prior to enrolment. It seems slightly misleading to describe the change in captopril-treated patients as an 'improvement', when a substantial proportion of them had been previously treated with other drugs, which themselves may have caused side-effects. Nonetheless, the unqualified claim that ACE-inhibitors such as captopril 'improve' patient well-being has been repeated in advertisements and journal articles ever since, seeming to imply that an asymptomatic individual with hitherto undetected hypertension could positively benefit from a dose of the drug.

Following the captopril study, a more blatant claim was made by the manufacturers of enalapril, a rival ACE-inhibitor (Cooper et al., 1987). In this open (i.e., uncontrolled), post-marketing surveillance of hypertensive patients in general practice, the most frequently recorded event during enalapril therapy was reported to be 'an improvement in well-being' (19.8 per cent). A close reading of the Methods section again reveals that 18.9 per cent of the study population had been commenced on enalapril following adverse effects from previous antihypertensive therapy. In addition, each patient was followed up on average for only 1.8 months; it is almost certain that further adverse effects would have been reported over a longer follow-up period.

'There is,' according to Oliver, 'a world of difference between removing a risk factor, such as cigarette smoking or oral contraceptives, and adding an unknown one, such as a drug' (Oliver, 1982). Yet the level of financial investment in trials of drugs that 'improve' the quality of life dwarfs the funding available for research on such fundamental questions as: the possibility of reducing the dosage of drugs, or withdrawing them altogether, following a period of treatment (Hudson, 1988); the feasibility of non-pharmacological approaches to reduce blood pressure (Kopelman and Dzau, 1985); and the elicitation of patients' preferences for antihypertensive treatment once informed of the benefits and risks (Shapiro et al., 1987). Pharmaceutical companies are understandably reluctant to fund studies which are unlikely to bring profits (or indeed actually to diminish the demand for their products). The Health Department and other funding bodies have tended, in their turn, to abrogate the responsibility of funding clinical research to drug companies. The resultant paucity of clinical trial evidence to justify non-pharmacological management — such as salt reduction and other 'lifestyle' interventions — has led to the tendency for such alternatives to get short shrift from physicians.

The situation is set up for a vicious cycle in which the 'nihilistic attitude' of physicians towards non-pharmacological management propels a further resort to drug treatment. Indeed, it has been said that:

The preference for a medical strategy of prevention over social strategies or individual behavioral approaches is partly a consequence of the availability of antihypertensive medication and the disposition of modern medicine to over-employ its wares. Not only does this disposition deny patients other options; it builds into the lives of hypertensive individuals a dependence on medication and regular interaction with the health care system (Guttmacher et al., 1981).

Over-medication manifests itself in other ways as well. For example, there has been a gradual increase in the number of so-called fixed-dose combinations entering the antihypertensive drug market. Diuretics, such as hydrochlorothiazide, are being combined with other classes of drugs in fixed dosages and sold on the market. A major problem with these formulations is that they deny the prescriber the choice of titrating the optimal dose for individual patients. Many drugs, including hydrochlorothiazide, exhibit a so-called 'flat dose response' whereby the blood pressure-lowering effect of the drug cannot be increased beyond a certain dose. In the case of hydrochlorothiazide, the threshold dose at which this property is observed is 12.5 milligrams (Berglund and Andersson, 1976). However, many fixed-combinations containing hydrochlorothiazide are currently marketed at doses as high as 25 milligrams, i.e., twice the minimum effective dose. This marketing strategy has sometimes been defended on the grounds that it ensures a maximal response nearly every time the pill is prescribed (Millar and Waal-Manning, 1990).

Unfortunately, this view ignores the drawback of such fixed-dose preparations, namely that the risks of many adverse effects are linearly related to the dose; in other words, there is no flat dose response for side-effects.

Perhaps the most extreme manifestation of the current medical dependence on drug treatment is described in the following report:

> The availability of a number of different classes of [antihypertensive] agents, usually with more than one drug per class, has created a new situation: the patient who reacts to virtually every drug prescribed . . . More aggressive treatment and the existence of a wide variety of generally well-tolerated drugs has led to the evolution of a new phenomenon: the 'I-never-took-pills-before' syndrome [in which the patient reacts adversely to virtually every pill prescribed] . . . The typical individual who manifests this syndrome readily admits to a dislike of taking medication in general . . . The admission of problems with tablet taking may signify a major attitudinal problem of the patient (Myers, 1989).

Nowhere in this monograph is the possibility entertained that the patient may be reluctant to take medication for a valid reason, or that the adverse effects experienced may be real. Labelling the patient with a new syndrome is apparently the medical profession's way of coping with their own drug dependence.

A Prescription for Change?

The current approach to the management of hypertension is the end result of the interactions between the medical profession and the pharmaceutical industry which have occurred throughout the evolution of our understanding of this problem. The balance of vested commercial and professional interests in mass treatment are unlikely to change in the near future. Once an innovation has successfully attained the status of a standard procedure, it proves difficult in many instances to reverse the process, particularly when there is a perceived absence of a viable alternative to replace it (Kawachi and Wilson, 1990).

A detailed prescription for change is beyond the scope of this essay. However, the options for reform are the same as those available in other instances of technological over-use: firstly, to develop more effective guidance and funding criteria for clinical research; secondly, to change medical education to produce more discerning users of medical technology; thirdly, to develop better information on efficacy, costs, and social effects of medical technologies, and to disseminate it to providers and users; fourthly, to strengthen regulatory programmes; and lastly, to use the financing system more aggressively to make the use of medical technologies more rational (Banta et al., 1981).

References

Amery, A., Birkenhager, W., Brixko, P., Bulpitt, C. et al. (1985) Mortality and morbidity results from the European Working Party on High Blood Pressure in the Elderly (EWPHE) trial *Lancet* ii: 1349-54.

Anonymous (1972) History of a hypertensive *Lancet* ii: 1243-4.

Australian National Blood Pressure Study Management Committee (1980) The Australian therapeutic trial in mild hypertension *Lancet* i: 1261-7.

Banta, H.D., Behney, C., and Williams, J.B. (1981) *Toward Rational Technology in Medicine* New York, Springer & Co.

Beaglehole, R. (1986) Medical management and the decline in mortality from coronary heart disease *British Medical Journal* 292: 33-5.

Beaglehole, R., Bonita, R., Jackson, R., and Stewart, A. (1988) Prevention and control of hypertension in New Zealand: a reappraisal *New Zealand Medical Journal* 101: 480-2.

Beaglehole, R., Dobson, A., Hobbs, M.S.T., Jackson, R., and Martin, C.A. (1989) Coronary heart disease in Australia and New Zealand *International Journal of Epidemiology* 18 (Supplement 1): S145-8.

Berglund, G. and Andersson, O. (1976) Low dose hydrochlorothiazide in hypertension: Antihypertensive and metabolic effects *European Journal of Clinical Pharmacology* 10: 177-82.

Berglund, G. (1984) Are we overtreating hypertension? *Acta Medica Scandinavica* 216: 337-9.

Bonita, R. and Beaglehole, R. (1982) Trends in cerebrovascular disease mortality in New Zealand *New Zealand Medical Journal* 95: 411-14.

Bonita, R. and Beaglehole, R. (1986) Does treatment of hypertension explain the decline in mortality from stroke? *British Medical Journal* 292: 191-2.

Breckenridge, A. (1985) Treating mild hypertension *British Medical Journal* 291: 89-90.

Brett, A.S. (1984) Ethical issues in risk factor intervention *American Journal of Medicine* 76: 557-61.

British Hypertension Society Working Party (1989) Treating mild hypertension: Agreement from the large trials *British Medical Journal* 298: 694-8.

Cooper, W.D, Sheldon, D., Brown, D., Kimber, G.R., Isitt, V.L., and Currie, W.J.C. (1987) Post-marketing surveillance of enalapril: experience in 11710 hypertensive patients in general practice *Journal of the Royal College of General Practice* 37: 346-9.

Croog, S.H., Levine, S., Testa, M.A. et al. (1986) The effects of antihypertensive therapy on the quality of life *New England Journal of Medicine* 314: 1657-64.

Dollery, C. (1987) Hypertension *British Heart Journal* 58: 179-84.

Forrow, L., Wartman, S.A., and Brock, D.W. (1988) Science, ethics, and the making of clinical decisions. Implications for risk factor intervention *Journal of the American Medical Association* 259: 3161-7.

Goldman, L. and Cook, F.E. (1984) The decline in ischemic heart disease mortality rates: an analysis of the comparative effects of medical interventions and changes in lifestyle *Annals of Internal Medicine* 101: 825-36.

Gross, T.P., Wise, R.P., and Knapp, D.E. (1989) Antihypertensive drug use. Trends in the United States from 1973 to 1985 *Hypertension* 13 (Supplement I): I-113-18.

Guttmacher, S., Teitelman, M., Chapin, G., Garbowski, G., and Schnall, P. (February 1981) Ethics and preventive medicine: the case of borderline hypertension *Hastings Centre Report* 11 (1): 12-20.

Hampton, J.R. (1986) Evidence suggesting that mild hypertension need not be treated *Journal of Hypertension* 4 (Supplement 5): S528-32.

Havlik, R.J., LaCroix, A.Z., Kleinman, J.C., Ingram, D.D., Harris, T., and Cornoni-Huntley, J. (1989) Antihypertensive drug therapy and survival by treatment status in a national survey. *Hypertension* 13 (Supplement 1): I-28-32.

Hennekens, C.H. (1987) Benefits of treatment of mild to moderate hypertension *Journal of General Internal Medicine* 2: 438-41.

Hudson, M.F. (1988) How often can antihypertensive therapy be discontinued? *Journal of Human Hypertension* 2: 65-9.

Jackson, R.T., Beaglehole, R., and Stewart, A.W. (1983) Blood pressure levels and the treatment of hypertension in Auckland, 1982 *New Zealand Medical Journal* 96: 751-4.

Joint National Committee on Detection, Evaluation, and Treatment of High Blood Pressure (1984) The 1984 report *Archives of Internal Medicine* 144: 1045-57.

Kaplan, N.M. (1990) *Clinical Hypertension* Baltimore, Maryland, Williams & Wilkins, fifth edition.

Kawachi, I., Malcolm, L.A., and Purdie, G. (1989) Variability in antihypertensive drug therapy in general practice: results from a random national survey *New Zealand Medical Journal* 102: 307-9.

Kawachi, I. and Malcolm, L.A. (1989) The rising expenditure on antihypertensive drugs in New Zealand, 1981-1987 *Health Policy* 12: 275-84.

Kawachi, I. and Wilson, N.A. (1990) The evolution of antihypertensive therapy *Social Science and Medicine* 31: 1239-43.

Klag, M.J., Whelton, P.K., and Seidler, A.J. (1989) Decline in US stroke mortality. Demographic trends and antihypertensive treatment *Stroke* 20: 14-21.

Kopelman, R.I. and Dzau, V.J. (1985) Trends in the therapy for mild hypertension. A word of caution *Archives of Internal Medicine* 145: 47-9.

Levinson, P.D. and Carleton, R.A. (1987) Mild hypertension and public policy. A perspective *Journal of General Internal Medicine* 2: 444-6.

Levy, R. and Moskowitz, J. (1982) Cardiovascular research: decades of progress; a decade of promise *Science* 217: 121-9.

Malcolm, L.A. (1990) Economic factors in formulating a national policy on the prevention by drugs of cardiovascular disease *New Zealand Medical Journal* 103: 402-4.

McKinlay, J.B. (1981) From 'promising report' to 'standard procedure': seven stages in the career of a medical innovation *Millbank Memorial Fund Quarterly* 59: 374-411.

Medical Research Council (MRC) Working Party (1985) Medical Research Council trial of treatment of mild hypertension: principal results *British Medical Journal* 291: 97-104.

Memorandum from the WHO/ISH (1986) 1986 guidelines for the treatment of mild hypertension *Hypertension* 8: 957-61.

Miall, W.E. (1986) The mild hypertension dilemma: Results of the British MRC trial *Journal of Clinical Hypertension* 3: 12S-21S.

Millar, J.A. and Waal-Manning, H.J. (1990) Minimum effective dosage in the drug treatment of hypertension *New Zealand Medical Journal* 102: 304.

Myers, M.G. (1989) The 'I-never-took-the-pills-before' syndrome and the treatment of hypertension *Canadian Family Physician* 35: 65-7.

Oliver, M.F. (1982) Risks of correcting the risks of coronary disease and stroke with drugs *New England Journal of Medicine* 306: 297-8.

Restall, P.A. and Smirk, F.H. (1950) The treatment of high blood pressure with hexamethonium oxide *New Zealand Medical Journal* 49: 206-9.

Shapiro, A.P., Alderman, M.H., and Clarkson, T.B. et al. (1987) Behavioural consequences of hypertension and antihypertensive therapy *Circulation* 76 (Supplement 1): I-101-3.

Simpson, F.O. (1990) Managing hypertension: drugs, life-style manipulation or benign neglect? Medical, ethical and economic considerations *Australia and New Zealand Medical Journal* 20: 726-34.

Sleight, P. (1985) High blood pressure: What level to treat? *Journal of Cardiovascular Pharmacology* 7 (Supplement 1): 109-11.

Stamler, J. (1978) The mass treatment of hypertensive disease: defining the problem *Annals of the New York Academy of Science* 304: 333-58.

Thom, T.J. (1989) International mortality from heart disease: rates and trends *International Journal of Epidemiology* 18 (Supplement 1): S20-28.

Thomson, G.E., Alderman, M.H., Wassertheil-Smoller, S., Rafter, J.G., and Samet, R. (1981) High blood pressure diagnosis and treatment: consensus recommendations vs actual practice *American Journal of Public Health* 71: 413-16.

Veterans Administration Cooperative Study Group (1970) Effects of treatment on morbidity in hypertension II. Results in patients with diastolic blood pressure averaging 90 through 114 mmHg *Journal of the American Medical Association* 213: 1143-52.

Wenger, N.K. (ed) (1984) *Quality of Life and Cardiovascular Care* New York, Le Jacq.

Wilcox, R.G., Mitchell, J.R.A., and Hampton, J.R. (1986) Treatment of high blood pressure: should clinical practice be based on results of clinical trials? *British Medical Journal* 293: 433-7.

The Exploitation of Fear: Hormone Replacement Therapy and the Menopausal Woman

Sandra Coney

The prospect of treating large populations of well people offers irresistible bounty to the pharmaceutical industry. If doctors and the public can be convinced that continuing good health is conditional on the use of drug products, the benefits to the industry are immense.

For all — Forever

The promotion of hormones to menopausal women is a unique phenomenon. Women are urged to take oestrogen to suppress the symptoms of menopause and to forestall various diseases they may develop as they age. No other group in society has been targeted for the universal prophylactic use of drugs. In all other cases, an indication must exist in the individual before a drug is prescribed. Anti-hypertensives are one of the more commonly used types of drugs in the pharmacopoeia, and it could be argued that hypertension is a relatively common condition, especially amongst older people. Yet even here, and in spite of considerable pharmaceutical company activity, treatment is reserved for those with demonstrably raised blood pressure, and there is considerable argument in the medical world as to the level of hypertension at which treatment might be commenced (see Ichiro Kawachi's chapter).

The only prerequisites leading to a recommendation to use hormone replacement therapy (HRT) are female gender and being in the age group approaching menopause. Women with no symptoms and no firm indications for intervention are receiving long-term treatment with powerful hormonal steroids. This situation has been made possible by defining the climacteric and the postmenopause — a period of women's lives stretching from the mid-forties until death — as an endocrine deficiency or disease state which can be cured by drugs.

By the late 1980s it was being argued in some medical quarters that most, if not all, menopausal women should be using hormones (Lobo and Whitehead, 1989; Mishell, 1989). Doctors expressed concern that the use of menopausal hormones in the population was too low (Barlow, Brockie, and Rees, 1991), and some studies outlined strategies for increasing their use (Ferguson, Hoegh, and Johnson, 1989). One study reported on strategies employed by a general practice which were successful in increasing the use of HRT among 40- to 60-year-old clients from 15 to 45 per cent (Coope and Roberts, 1990). Typically, one New Zealand doctor said that:

Currently in New Zealand fewer that 10 per cent of post-menopausal women use HRT. If anything, the menopause for New Zealand women is under medicalised and many women are unaware of the benefits that HRT may give them (Farquhar, 1991).

It was increasingly proposed that doctors should actively recruit users and that women should use hormones from menopause to the end of life. The consensus of the 13th Congress of FIGO (International Federation of Gynaecology and Obstetrics) held in September 1991 was that:

Doctors should offer hormone replacement therapy (HRT) to all menopausal women to use indefinitely, irrespective of whether they have menopausal symptoms . . . delegates voted almost unanimously for the routine use of HRT. They also supported the proposal that the treatment should be continued indefinitely rather than for the maximum 10 years currently recommended (Barnes, 1991).

The same message was being promoted to women through the mass media. Typically, the medical adviser in a major women's magazine said in answer to a woman's query about how long hot flushes lasted that:

You must understand you need hormone replacement . . . whether you have hot flushes or not . . . Only in exceptional situations can menopausal women not take hormones (Dr Dan, 1991).

Studies showed that many women had internalized the message that hormones were routinely necessary in the menopause. They often initiated treatment themselves (Hunt, 1988) and women using hormones had no more severe menopausal symptoms than women who did not (Ferguson, Hoegh, and Johnson, 1989).

To understand how this situation has evolved requires an examination not just of the marketing tactics of drug companies, but of the interrelationship between the medical profession and the pharmaceutical industry, as well as the methods used to exploit negative social stereotypes of older women.

Laying the Ground

Certain preconditions needed to exist before well women could be persuaded that they would benefit from the long-term use of drugs. These included:

- Women's existing relationship with the health system which allowed the medicalization of a normal life event.
- Anxiety among women about the diminution of their social status as they aged.
- The re-definition of the normal biological event of menopause as a disease state which could be treated.
- Appropriation of serious non-gender-specific diseases for inclusion in the 'new menopause'.
- A 'campaign' created through tacit co-operation between the pharmaceutical industry, medical opinion leaders, and the media.

Women: A Suitable Case for Treatment

Women especially have been the targets of campaigns to induce healthy people to use drugs and they are uniquely vulnerable to the pressure to comply. There are several reasons for this. Firstly, it is relatively easy to define women as in need of treatment. Medicine is inherently sexist in that it proceeds from the assumption that the male is the norm (Howell, 1974; Scully and Bart, 1973). Humanness is assumed to be male, and the female is simply subsumed into this paradigm. For example, research has often been conducted on male populations, and the results then crudely extrapolated to women. Through this distorted lens, the biological functions of women can be readily categorized as abnormal or pathological. Women's relationship with the health system is bedevilled by the tendency to consign physical symptoms reported by women to psychological causes such as a failure to accept the female role, infantilism, hysteria or neuroticism (Lennane and Lennane, 1973; Polivy, 1974). Women are more likely to be labelled as having emotional problems by physicians even when reporting identical symptoms to men (McKintyre and Oldman, 1977). This gender bias provides the ideological justification for medical surveillance of women's lives, and for the medical treatment of healthy women.

Secondly, women's experience of the health system conditions them to accept the medicalization of normal life events. Women make greater use of health services, and unlike men, must use them even when they are well, for purposes such as the control of fertility, abortion, pregnancy, and childbirth (Kaufert, 1980). Immunization against rubella, and attendance at medical clinics for immunization of their children, cervical screening, and mammographic screening, are other examples of the way women have regular contact with health care providers for routine check-ups and preventive interventions.

As a result, dependence on the health system becomes normalized for women. They accept medical oversight of their lives, especially those aspects of their lives which are exclusive to their sex.

Thirdly, the relationships of authority and submission women encounter in the health services are congruent with the power imbalance in gender roles in society at large and thus go largely unchallenged (Bell, 1987). Women tend to believe that the health care they are offered is 'for their own good'. They usually fail to perceive the sexist underpinnings of medicine and how this affects what they are offered. Consequently, they constitute a relatively unquestioning and receptive population when it comes to campaigns to market interventions in their lives. This works to the advantage of the pharmaceutical industry and is in turn exploited by it.

There is an added feature of the female population which is open to exploitation by the pharmaceutical industry. The social status of women in modern Western societies is inferior to that of men, and success for women is still partially measured in terms of attractiveness, youth, and sexual desirability. Even the woman who is in every other way successful in the fields of commerce, sport or politics may still be judged as deficient if she diverges from the female stereotype. Fatness, signs of normal ageing, and variations in physical appearance are viewed as problems which can be cured medically.

Conversely, women who are unable to happily conform to the social role expected of them as females may suffer mental trauma. Health care providers have traditionally regarded these socially-caused manifestations of distress as medical problems which can be cured with medication. They rarely challenge the value system of the society from which the problems arise. The pharmaceutical industry has reinforced this behaviour by its marketing strategies and has exploited stereotypes of gender roles in specific campaigns. The epidemic levels reached in the prescribing of benzodiazepams to depressed suburban housewives in the 1960s and 1970s provides an example of this, as does the campaign to promote the use of amphetamines and other psycho-active drugs by dieting women in the same period. In the 1990s, the promise held out for drugs such as Retin-A and HRT is that they will preserve youthful femininity, and in the case of HRT, that it will cure a whole host of psychological symptoms, have a 'mental tonic effect', and induce a feeling of well-being.

The targeting of well, mid-life women for mass medication with hormones provides a case study in the strategies of the drug companies to create and expand markets.

Women and Ageing

Fear of ageing has cultural causes. The low status given to older women in many industrialized societies profoundly affects women's attitude towards growing older.

The widespread use of menopausal hormones is largely confined to developed countries. HRT is a treatment of affluence, and even within such countries, use rates are highest among women in higher socio-economic groups. When the association between oestrogen use and the risk of endometrial cancer was established in the mid-1970s, an assistant commissioner of the United States (US) Food and Drug Administration (FDA) commented that:

[T]he chronic users [of oestrogen] tend to be middle class and upper income women, the kind of people who go to doctors . . . It is an interesting example of the poor being spared (Anon., 1976).

Later studies have demonstrated that the same class pattern of use has persisted. A British study found that 46 per cent of hormone users were in social classes one and two compared to a corresponding census figure of 23 per cent (Hunt et al., 1987), and a Swedish study found that higher education led to 'a more than two-fold increase in the risk of receiving oestrogen treatment' (Bergkvist et al., 1988).

Western capitalist societies are marked by being highly competitive and socially stratified. Economic success determines social status, and the ideology of the marketplace requires that this be demonstrated, not just by visible wealth, but also by the possession of a healthy, beautiful body. For middle-class women appearance is one of their prime assets. The body provides a visual display of success; it is an emblem of the woman's status in the world.

[W]omen's concern with their appearance is not simply geared to arousing desire in men. It also aims at fabricating a certain image by which, as a more indirect way of arousing desire, women state their value. A woman's value lies in the way she *represents* herself . . . (Sontag, 1974).

To look fit, slim, and attractive are goals which are sought by middle-class women in gyms, health food shops, beauty parlours, and doctors' surgeries throughout the Western world. The latest manifestation of this quest is the cosmetic surgery clinic. The normal process of ageing is anathema to this pursuit.

Only one standard of female beauty is sanctioned: *the girl* . . . The single standard of beauty for women dictates that they must go on having clear skin. Every wrinkle, every line, every grey hair, is a defeat. No wonder that no boy minds becoming a man, while even the passage from girlhood to early womanhood is experienced by many women as their downfall, for all women are trained to want to continue looking like girls (Sontag, 1974).

In Western societies, ageing for women is a 'process of gradual sexual disqualification' (Sontag, 1974) and increasing social invisibility. Men gain status as they age; women lose it.

Sooner or later the middle-aged woman becomes aware of a change in the attitude of other people towards her. She can no longer trade on her appearance, something which she has done unconsciously all her life. There is no defined role for her in modern society . . . (Greer, 1991).

Women are encouraged to fight this process of normal ageing to maintain their status. The media, and in particular women's magazines, are the principal purveyor of the view that women have a duty to stay young by any means (Coney, 1991; Wolfe, 1991). Women are exposed to a relentless message that ageing in women is unacceptable and that a woman must, as Germaine Greer expresses it, 'masquerade as a girl to remain in the land of the living' (Greer, 1991).

Few women are immune to these messages. They touch deep fears, threatening a woman's sense of personal identity and self-esteem. As she ages, a woman loses the twin abilities to reproduce and attract admiration for her appearance. The loss makes her vulnerable to exploitation by commercial interests which promise the preservation of youth.

The conceptual meaning of HRT is of stopping the clock, of making time stand still. HRT is a denial of old age, a tool women can employ to attempt to outwit natural processes. Women using hormones are in a kind of limbo: unable to reproduce, but still menstruating, and with their drugs providing what their ovaries can no longer produce. The intrinsic promise of hormones is of maintenance of youth.

The promotion of hormones has involved the suggestion to women that the drugs will enable them to retain their youthful femininity by preventing the so-called 'dowager's hump', skin ageing, and the thinning of hair. In addition, the artificial maintenance of menstrual periods obscures the actual loss of reproductive function, enabling women to delude themselves into 'feeling young' despite the passage of the years.

The scientific evidence for an anti-ageing effect of oestrogen is paltry. Most studies have shown no improvement in the skin of hormone users. Skin ageing is related to the general ageing process, and sun exposure and cigarette smoking have been pin-pointed as the major culprits (Kadunce et al., 1991). If oestrogen does plump out wrinkles it is probably through fluid retention, an effect of oestrogen which also leads to uncomfortable bloating and headaches.

Overt encouragement of women to use possibly cancer-causing drugs to retain a youthful appearance is an ethically perilous position and could undermine the credibility of any pharmaceutical marketing campaign. Consequently, the strategies adopted in encouraging 'youth drug' status for HRT have been more subtle.

The official position of the pharmaceutical companies is spelled out in a patient insert for the oral oestrogen Premarin. In this, Ayerst Laboratories stated that though women

> may have heard that taking estrogens for long periods (years) after the menopause will keep your skin soft and supple and keep you feeling young [t]here is no evidence that this is so . . . (Ayerst Laboratories, n.d.(a)).

Nevertheless, much of the advertising to doctors through medical journals has indirectly conveyed the message that hormone users will remain youthful. This is sometimes achieved by the use of models who are clearly less than menopausal age (Coney, 1991) or slogans such as 'For more years of femininity and joie de vivre' (Schering, 1975) and 'So a woman can continue to enjoy being a woman' (Schering, 1991).

A booklet for patients on menopause and HRT produced by Ayerst 'as part of a caring commitment to female health' does not baldly claim any youth-preserving properties for HRT, but makes the suggestion in other ways. '[W]eight gain, as well as hair and skin changes' are listed as symptoms of menopause (Ayerst Laboratories, n.d.(b)), the implication being that these will be cured along with other symptoms by the use of hormones.

The message through the lay media is more explicit. 'Hormone replacement therapy', said the women's magazine *Woman's Day*, makes many women 'feel younger and more beautiful . . . When cells have the oestrogen they need, the facial skin and breasts have a healthy, fleshy appearance' (Donaldson, 1991). Women are also told they need oestrogen to maintain their libido and have comfortable sex.

The Amarant Trust, a British group promoting the use of HRT to women, is named after the Amarant, a mythical, never-fading flower, a symbol of immortality and enduring beauty to the Greeks. The Trust has paraded a succession of public figures, claiming that their good looks and youthful appearance can be put down to the hormones they take. Some of the work of the Amarant Trust is supported by grants from manufacturers of hormones, and a handbook for women published by the Trust is co-written by Dr Malcolm Whitehead (Gorman and Whitehead, 1989), a British doctor who features prominently in promotional material for oestrogen patches produced by Ciba-Geigy.

Among packages handed to New Zealand physicians by drug detailers are papers giving favourable results from skin tests on oestrogen users. A general practitioner newsletter, *Health Update*, sent by a New Zealand general practitioner to his clients promised that oestrogen not only 'stops . . . hunchback' but 'delays skin ageing process' (Bonita, 1990). The promise of prolonged youthfulness through the use of hormones is an especially powerful strategy for persuading women to use them.

The power of this promise for women is illustrated by the following description of one woman's decision to use oestrogen. A breast cancer patient, she nevertheless sought treatment for hot flushes and for other reasons:

I didn't want to atrophy, have a dry vagina, get osteoporosis, be a crone . . . Now I can be nine years old, or eighteen, or early-married, or a young acrobat, or even a cranky crone. I don't have to be menopausal, a woman 'in menopause', a woman entering a new time of life, a woman now free to enjoy sex without fear. I'm all ages, all pasts, all futures, as long as I take my beautiful pill each morning (Lawrence, 1991).

Menopause as Disease

The recommendation that menopausal hormones should be universally used depended on a change in the meaning of menopause.

Until the early 1960s, menopause was largely viewed by medicine as an event in women's lives with psychological rather than physical causes and effects. In the decades from the 1920s to the 1950s the work of Freudian psychoanalysts such as Therese Benedek (1950) and Helene Deutsch (1945) were influential. Menopause was depicted by them as obsolescence; woman had 'reached her natural end — her partial death — as servant of the species' said Deutsch. By confining woman to a role of sexual and reproductive being, the argument could be made that menopause must be experienced by women as a trauma, thus explaining the apparent emotional lability of women at this time. The later definition of menopause as an oestrogen-deficiency disease built on and incorporated this psychological theory.

The genesis of the disease orientation towards menopause lay in the 1930s and 1940s with the work of research scientists in the field of sex endocrinology (Bell, 1987). Through their scientific work, endocrinologists were able to demonstrate the mechanism of ovarian function and the drop in hormone levels at the menopause. The re-definition of menopause as an 'oestrogen deficiency disease' went hand in hand with the invention of various artificial sex hormones during the same period.

In the 1930s and 1940s, scientists finally succeeded in developing commercially viable synthetic oestrogens: oestradiol in 1933, diethylstilboestrol or DES in 1938, ethinyloestradiol in 1938, and conjugated equine oestrogen (brand name Premarin) in 1943.

The availability of these drugs changed the status of menopause. It was no longer of little interest to physicians, who had formerly abandoned menopause to women themselves or treated them with psychotropic drugs, but a prime area for medical 'empire-building' (Greer, 1991). Howard Judd and Wulf Utian, two doctors prominent in the menopause field, said it was the development of these oestrogenic preparations that changed the medical opinion of menopause from that of 'a side issue of nuisance complaints to an issue central to the subject of aging in women' (Judd and Utian, 1987).

The commercial potential of these discoveries could not immediately be realized. Many doctors were suspicious of sex hormones, as there had been reports of oestrogens causing cancer in laboratory animals since the 1930s. Medical specialists were also ambivalent about the applicability of scientific findings to their clinical work. Underlying this was a fear that laboratory science could diminish medicine's power, undermining the doctor's clinical judgement in individual cases (Bell, 1987).

Sociologists have argued that the process of medicalization is assisted when new conceptual ideas are championed by 'elite' opinion leaders in a profession (Conrad and Schneider, 1980). In the case of the re-definition of menopause, the 'champion' was Dr Robert Wilson, a prominent New York gynaecologist with a large practice. As a practising clinician, he was able to overcome the reservations of many of his fellow practitioners, by describing in detail the use of oestrogen on his own clients.

It was Wilson who contributed most to the elaboration and establishment of menopause as an oestrogen-deficiency disease. His campaign was not concerned with anything as trivial as the alleviation of menopausal symptoms such as hot flushes. He argued that once women reached menopause and their levels of oestrogen declined, they ceased to be real women. 'The unpalatable truth must be faced', he said, 'that all postmenopausal women are castrates' (Wilson and Wilson, 1963).

Postmenopausal women were 'desexed' and 'defeminized' by the loss of oestrogen. '[Oe]strogen-starved women', Wilson said, suffered from an immense range of physical and psychological ailments including 'a vapid cow-like feeling called a negative state', osteoporosis, tough, dry skin, flabby breasts, and vaginal atrophy. According to Wilson:

... The more intelligent woman instinctively knows that her loss of physical attractiveness is entirely out of proportion. She sees the marked skin changes, the disfiguring fat deposits, the atrophy of her breasts and the beginning disappearance of her external genitals ... All this has a profound effect upon her psyche (Wilson and Wilson, 1963).

Wilson proposed the elimination of the menopause by the use of oestrogen, and like his modern-day counterparts, he saw no reason for women to ever stop taking it. 'A beneficial estrogen level should be continued throughout life', he said, and oestrogen should be present in the female 'from puberty to the grave' (Wilson and Wilson, 1963).

Dr Wilson wrote medical papers to persuade his fellow professionals to adopt his theories and clinical practices. Many doctors initially looked askance at him because his paean of praise on the benefits of oestrogen seemed extravagant and his methods entrepreneurial, but his ideas were taken up by eminent doctors such as Dr Robert Greenblatt and Dr Wulf Utian, adding weight and respectability to the pro-hormone message.

Wilson was also generously supported by drug companies, as were many of the hormone apostles who were to follow him. He established the Wilson Foundation in 1963 for the purpose of promoting oestrogen with $US1.3 million from the pharmaceutical industry (Seaman and Seaman, 1977). In one year (1964) Wilson received $US17,000 from G.D. Searle Ltd, $US8700 from Ayerst Laboratories and $US5600 from the Upjohn Company. All these companies made hormones which Wilson promoted as effective in the treatment of menopause. It was a mutually beneficial relationship, for the pharmaceutical companies were able to replicate Dr Wilson's arguments and even his graphs in their advertisements in medical magazines and journals.

Wilson also took his message to the lay public. His book, *Feminine Forever* (Wilson, 1966) sold 100,000 copies in the first seven months after publication, and was excerpted widely in women's magazines such as

Vogue. A survey conducted in 1969 by the International Health Foundation, a drug company-funded establishment, looked at women's knowledge of the use of HRT in each of five European countries. It concluded that the differences in knowledge were entirely explainable by the availability of Wilson's book, and the extent of publicity surrounding it. In West Germany, leading experts had quickly endorsed Wilson's views and there was widespread publicity. The survey found that 71 per cent of West German women knew about hormone treatment, as did 54 per cent of women in France, 47 per cent of women in Italy, and 36 per cent of women in Britain (Van Keep, 1990).

Wilson's ideas, although judged by some at the time as extreme, have become the orthodox medical view of menopause in the 1990s. In 1990, Dr Pieter van Keep of the International Menopause Society, said that for him HRT began when he read the 1963 paper by Robert Wilson (Van Keep, 1990). Many of his ideas and even his language are repeated with all the authority of absolute scientific truth. Dr Wilson's targeting of potential users of oestrogen through the lay media has also continued as a favoured tactic of the promoters of oestrogen. Women are bombarded with injunctions to use oestrogen to prevent the ravages of old age. Numerous books have been produced for women, often co-written or endorsed by prominent doctors (Cooper, 1975; Nachtigall and Heilman, 1986; Gorman and Whitehead 1989). Women's magazines eulogize about the benefits of oestrogen under headings such as 'Oestrogen — the menopause miracle . . . the answer to your change of life nightmares', quoting doctors who believe that 'oestrogen is of vital importance . . . [and] necessary for happiness, sexuality and physical well-being' (Donaldson, 1991).

The Ideology of the 'New' Menopause

The ideology underlying the argument for the menopause as an oestrogen-deficiency state is deeply influenced by cultural attitudes about the proper roles of women and by negative stereotypes of older women. The concept that oestrogen is necessary to femininity in essence reduces woman to her hormones:

Integral to the medical model is the equating of women's biological role with the meaning of her existence. Through this perspective of biological determinism, woman is her hormones. Her life is defined by her fertility. Her most 'normal' period and the apogee of her whole life, is her fertile period, between puberty and menopause . . . By this definition women have only a biological purpose in life and no social purpose at all. Their status is determined by the state of their ovaries (Coney, 1991).

The biomedical model also removes from women the ability to define the experience of menopause for themselves. Women can no longer confidently say they are menopausal; this can only be determined by a doctor after testing the level of ovarian hormones in the laboratory. Scientific objectivity has replaced the subjectivity of woman and woman is enjoined to seek medical assistance to determine whether she is menopausal, and to manage this period of her life.

Perversely, however, the medical model lacks science in critical respects. For example, one video for patients funded by the Upjohn Company and used by Ciba-Geigy, includes as a symptom of oestrogen deficiency 'feeling unloved' (MacLennan, 1988). The range of symptoms often given for the 'new' menopause includes a whole range of psychological and sexual symptoms such as loss of libido, depression, irritability, and lack of confidence, for which there is no scientific proof of a link with oestrogen (McKinlay, McKinlay, and Brambilla, 1987; Coney, 1991). The old ideas about women's emotional instability at this time of their lives have simply been appropriated by the medical model.

Numerous studies have shown that there is no increase in depression among menopausal women, and that men of the same age also experience many of the symptoms ascribed to women during menopause (Bungay, Vessey, and McPherson, 1980). Sexual activity declines slightly with age for women, but it also does for men. When studies are controlled for the presence of an available partner, there is no evidence that women's libidos decline at menopause (Youngs, 1990).

Many of the psychological symptoms blamed on menopause are the result of ageing, or are found in people of all ages and both sexes. There are social causes for many of these symptoms, such as stress from the social roles women play (Greene and Cooke, 1980; Cooke, 1985), but the proponents of hormones say they result from oestrogen deficiency and argue for treatment by hormones.

Defining menopause as a deficiency state or disease, and a medical matter, logically argued for its treatment:

The definition of menopause as a disease which can be 'cured' through estrogen replacement entails a series of moral obligations on both the physicians and patients. Given the therapeutic means to correct a condition defined as a disease and, therefore, 'bad', physicians are obliged by virtue of their calling to act, to treat the abnormal, the disease state. Reciprocally, people with a disease are under an obligation to seek treatment. Applied to menopause, the model implies that physicians are obliged to provide, and women to seek, estrogen therapy. Re-inforcing the obligation to restore the individual to normality, there is also a socio-economic obligation to the wider community (Kaufert and Gilbert, 1986).

According to Dr Robert Lindsay, a prominent researcher in the area of osteoporosis:

There should not even be debate on this issue [the use of HRT]. It is almost criminal not to make HRT available to all post menopausal women (Barnes, 1991).

Medical symposiums and articles in the lay media stress the cost to nations and communities of the morbidity which, it is claimed, arises from the 'untreated' menopause. Frightening figures are produced which purport to show the health care costs of osteoporotic fractures, most of which, it is argued, could be prevented by the widespread use of oestrogen.

Women not using hormones are now described as having 'untreated' menopauses (Utian, 1987). More recently the concept that menopause should be 'managed' has gained currency in medical circles (Anon., 1990). Typically, an Australian gynaecologist described menopause as 'a sex-linked endocrine deficiency disease which requires careful evaluation, management and follow-up' (Wren, 1987).

Menopause has now become a health hazard: 'Second only to smoking, oestrogen deficiency morbidity and mortality represents a largely preventable health hazard' (Ayerst Laboratories, 1991b).

The boundaries of menopause have expanded. Whereas the term formerly referred to a relatively brief period of women's lives, the 'deficiency' definition alters this time span. Menopause is now 'a permanent condition to be permanently managed' (Kaufert and Gilbert, 1986). This is of obvious benefit to the manufacturers of hormones. Menopause is described as a 'twentieth century disease' (Heine, 1991) because women are living longer, whereas in earlier times women died before they had many years in their oestrogen-deficient state:

... the climacteric is not a natural phenomenon, but a hormone deficiency state. Women (and men) now live longer than nature intended and women can consider themselves very fortunate that oestrogen deficiency can be easily rectified. Indeed, it is just as natural to give oestrogens to climacteric women as it is to give insulin to diabetic patients (Rauramo, 1986).

Since medicine has kept women alive beyond their reproductive lifespans, it is argued that medicine has a responsibility to make women's later years comfortable.

The word 'replacement' in the term 'hormone replacement therapy' implies that a normal condition is being regained. Hormone replacement therapy is depicted as more 'natural' than ageing. As one doctor put it:

> Hormone replacement therapy is a replacement — it's not a treatment . . . it is a method of making the 20 years that occur after menopause more natural. That's what we doctors are aiming at. To make life more natural (McKenzie, 1990).

According to an advertisement for one hormone product: 'With Ogen, you can give back what nature takes away. And because Ogen replicates a menopausal woman's natural oestrogen, it's a natural way of doing so' (Abbott Australasia Pty, 1991). Ayerst Laboratories puts its oestrogen in the same category as a product from a health food shop: 'Happily, Premarin Therapy is nearly as simple as taking Vitamin Supplements' (Ayerst Laboratories, 1991b). Ciba-Geigy, manufacturer of transdermal oestrogen patches, boasts that women regard their patches as 'little portable ovaries' (Adis International Ltd, 1990). The company claims that the medical consensus currently regards menopause not as a 'natural phenomenon' but as a hormone deficiency state that can easily be remedied: 'It is as natural to give estrogens to menopausal women as it is to give insulin to diabetic patients' (Ciba, 1990).

The indistinguishability of the medical and pharmaceutical positions on menopause justifies the confidence of the pharmaceutical industry in relying on physicians to recruit users of HRT.

The Appropriation of Major Diseases as Indications for HRT

The argument for the morbidity of menopause has been immensely enhanced by the appropriation of two serious diseases as consequences of menopause: osteoporosis and cardiovascular disease. Emphasizing the prevention of these diseases has been the primary marketing strategy of the makers of hormones since the mid-1980s.

In the 10 years following the first trumpet blast in Dr Wilson's campaign, dollar sales of oestrogen quadrupled in the US. By 1975, with prescriptions at an all-time high of 26.7 million, oestrogen was one of the top five prescription drugs in the US (McCrea, 1983). By this date, six million US women were regularly using oestrogen (Boston Women's Health Book Collective, 1984). One study in Seattle showed that 51 per cent of women over 50 years of age had tried oestrogen (Seaman and Seaman, 1977). Similar trends in sales, though not as dramatic, were seen in other countries. In Sweden, sales of oestrogen trebled from 1973 to 1977 (Persson et al., 1983).

This activity was enormously profitable for the drug companies. Ayerst's Premarin is the market leader in the US, with 75–80 per cent of oestrogen sales by the mid-1970s. In 1972, the total sales of American Home Products, Ayerst's parent company, were $US1587 million (Silverman and Lee, 1974); it was estimated that Premarin sales accounted for 10 per cent of this.

In 1975 new research 'shattered the hormone dreamtime' (Coney, 1991). The results of two studies, published in the *New England Journal of Medicine*, showed a vastly increased rate of endometrial cancers — between 4.5 and 14 times — among users of oestrogen (Smith et al., 1975; Ziel and Finkle, 1975). One of the studies showed that long-term users were at special risk (Ziel and Finkle, 1975). In the early days of oestrogen use doctors prescribed hormones for short periods only. Eventually, however, they succumbed to slogans encouraging indefinite use, such as Ayerst's 'Keep her on Premarin'.

'Estrogens forever', said one doctor a year before the crash, 'enjoys the status of a cult' (Allen, 1974).

Although the advocates of hormone use attempted to discredit the studies or claimed the cancer was of a less 'serious' kind (Cooper, 1975; 1978), the increase in cancer incidence was 'too high to be explained away by even major methodological flaws' (Kaufert and McKinlay, 1985).

Later, further studies showed that endometrial cancer rates had risen sharply in the US, parallel with the expansion in the use of oestrogens. In 1976, there was the first report of a higher breast cancer risk in users of oestrogen (Hoover et al., 1976), an even more alarming prospect than the endometrial cancer risk because of the higher incidence of breast cancer and the poor prognosis for even early-stage disease. The third risk raised by researchers during the same period was the report from the Boston Collaborative Drug Surveillance Program (1974) linking oestrogen with gallbladder disease. Another possibility suggested by researchers was that of an association of oestrogen with heart disease, whether beneficial or harmful.

The cumulative effect of these black marks against oestrogen was that sales plummeted. Prescriptions dropped sharply in Britain, and in the US oestrogen use declined by 18 per cent from 1975 to 1976, and by another 10 per cent from 1976 to 1977 (Kaufert and McKinlay, 1985). One factor in the decline was the insistence of the US FDA that packages of oestrogen preparations contain an insert warning of the risks. This was opposed by hormone manufacturers, the American College of Obstetrics and Gynaecology, the American College of Internal Medicine, and the

American Cancer Society. The College maintained that the insert violated clinicians' right to control how much information to disclose to patients, and threatened medicine's professional autonomy (Kaufert and McKinlay, 1985).

Despite the setback, Ayerst Laboratories barely paused in its promotional efforts with Premarin. 'Dear Doctor' letters were sent to physicians, and the addition of progesterone to the therapy was promoted as the way to protect the endometrium and avoid the endometrial cancer risk. Nevertheless:

> The endometrial cancer crisis created a need to find some additional reason to woo women onto hormones. Osteoporosis [thin bones] provided the answer: the condition was prevalent among women, and in serious cases was disfiguring and even deadly . . . the pharmaceutical promotional juggernaut swung into action (Coney, 1991).

In the mid-1980s, new research results showed that oestrogen modified bone loss, and, in 1986, US drug manufacturers obtained permission from the FDA to cite prevention of bone loss as an indication for the use of oestrogen. This provided an opportunity to re-market oestrogen for a much more serious purpose than allowed by the 'feminine forever' approach. In addition, as further study results came in, it became clear that long-term hormone use would be necessary to provide any protective effect on bones. Women were advised to use oestrogen for periods varying from six to 20 years to protect against osteoporosis, adding to the commercial potential of hormones.

Prior to the involvement of companies manufacturing hormones, osteoporosis had been seen as a bone disease, and was listed as such in medical textbooks (Van Keep, 1990). The condition affects both men and women as they age, although women have a higher incidence of fractures because they have less bone mass when bone loss commences, and they also experience an accelerated period of bone loss at menopause.

To succeed in deploying osteoporosis as an inducement to use hormones, it was necessary to re-orient perceptions of the disease. Osteoporosis had to be re-defined as a woman's problem and as a symptom of menopause. To persuade women of the benefits of oestrogen, women had to believe osteoporosis was their disease, and they had to believe it was caused by oestrogen deficiency. In addition, they had to perceive the cancer risks as trivial against the benefits of preventing bone loss.

Ayerst Laboratories hired the public relations firm of Burson-Marsteller to market osteoporosis. The purpose of the campaign, said a company vice-president, was 'to educate women about the existence of osteoporosis and risk factors associated with it' (Dejanikus, 1985). Women were to be targeted in two ways: directly, through the media and patient education, and indirectly, through health professionals. Stories on osteoporosis as a serious women's health problem were arranged on radio and television, and in magazines such as *Vogue*, *McCall's*, and *Reader's Digest*. The company's patient leaflets described osteoporosis as a menopausal problem and suggested that women discuss prevention with their doctors. Nurses were trained to take workshops on osteoporosis for church groups, women's clubs, and hospital outreach programmes.

Health care providers were influenced by other strategies. Three medical experts were toured through 10 major US cities. This tour capitalized on a consensus report on osteoporosis put out by the National Institutes of Health (NIH) in 1984. The NIH report had warned against the widespread use of hormones because of the unknown risks, but otherwise had stated that oestrogen was the 'most effective' way of preventing osteoporosis. Doctors were also courted through the medical media with a flood of drug company-inspired articles promoting the benefits of oestrogen.

Back in the 1960s, Dr Robert Wilson had used the image of the woman with the 'dowager's hump' to emphasize the relationship between oestrogen and postmenopausal bone loss. In the early years of promotion of hormones, Ayerst had also utilized the same image in its advertising, dubbing her 'the littler old lady' or the 'incredible shrinking woman'.

The symbols of the marketing campaign of the 1980s and the 1990s have represented the most crippling forms of osteoporosis: kyphosis (severe curvature of the spine) or hip fractures. A video for women, funded by drug companies, shows women shuffling painfully forward on walking frames while the voice-over discusses osteoporosis (MacLennan, 1988). In drug advertisements to doctors, canes, crutches, and wheelchairs have been featured with messages such as: 'Which would you rather prescribe to a potential osteoporosis sufferer? Take a course of Prempak-C? Or let nature take its course'. In patient literature, women are told that 'oestrogen deficiency . . . leads to loss of height in many women. A woman may lose an inch and a half for each decade after menopause. Over a lifetime, a woman may lose a total of 4 to 6 inches in height' (Ayerst Laboratories, n.d.(b)).

In cultures where women are valued for their appearance, the image of the deformed old woman — the crone — touches women's worst fears about ageing. The image is used not just by the makers of hormones, but by the makers of calcium supplements, and by the dairy industry, both of which have poured millions of dollars into campaigns to raise the profile of osteoporosis among the public (Coney, 1991). The effect of several concurrent campaigns has been cumulative.

These images are accompanied by statistics such as:

- 'The crippling disease that strikes 25% of all women' (Notelovitz and Ware, 1985).
- 'Many people have osteoporosis — 90% of these are postmenopausal women' (Ayerst Laboratories, n.d.(b)).
- '[Osteoporosis] leads to more deaths in older women than cancers of the cervix, ovaries and womb combined' (Cooper, 1990).

These claims distort the epidemiology of osteoporosis and exaggerate the likelihood that it will affect an individual woman. Statistics for the incidence of spinal fractures, for instance, have been derived from studies of chest X-rays taken of women entering hospital for reasons other than osteoporosis. The women concerned may have one or two fractures, but they may not even be aware that these have occurred. In the statistics quoted above, there is no differentiation between fractures causing health problems, and those which are not even known to the person. According to Dr Bruce Ettinger, a researcher in the osteoporosis field:

Women shouldn't worry about osteoporosis. The osteoporosis that causes pain and disability is a very rare disease. Only 5% to 7% of 70-year-olds will show vertebral collapse; only half of these will have two involved vertebrae; and perhaps one-fifth or one-sixth will have symptoms. I have a very big referral practice, and I have very few bent-over patients. There's been a tremendous hullaballoo lately, and there are a lot of worried women — and excessive testing and administration of medications (Pollner, 1985).

Assigning the cause of fractures, especially hip fractures, to oestrogen alone is clearly not accurate. The explanation ignores men, who also suffer hip and wrist fractures. Thin bones should be regarded as a risk factor for fractures, rather than the sole cause. Other factors include frailty, poor cognisance, disorientation, weak muscles, and reduced fat and muscles around the hip. Use of psychotropic and hypnotic drugs is common among elderly women, adding to the likelihood of falling.

A recent major review of strategies for the prevention of osteoporosis and fracture (Law, Wald, and Meade, 1991) concluded that hormone use would reduce hip fractures by 50 per cent only in comparatively young women who took oestrogen for decades. Hip fracture is uncommon in this age group. Even 20 years of use (from 50 to 70 years of age) would have little effect in women over 75 years of age in whom 80 per cent of all hip fractures occur.

Nevertheless, the perception among women who have access to the media hype, but not the fine print in medical journals, is that prevention of fractures is a major reason for considering hormones. A study of their women clients aged 50–52 years by a group of general practitioners in Britain produced the result that 80 per cent of the women described the most important problem of menopause as osteoporosis (Draper and Roland, 1990), while about three-quarters said they would be interested in taking hormones to prevent osteoporosis. An American study found that 88 per cent of women using HRT knew about risk factors for osteoporosis, compared to 28 per cent not taking it (Ferguson, Hoegh, and Johnson, 1989). Only 25 per cent in either group knew about the association between smoking and bone loss. This suggests that information given to women about osteoporosis has been presented in such a way as to encourage hormone use, rather than to help them identify modifiable risk factors in their lives.

Osteoporosis now figures prominently in any discussion of menopause in articles in women's magazines or other lay media. Osteoporosis is depicted as the principal symptom of menopause, with the most distressing outcome. It is also presented as entirely preventable by the use of hormones, despite the fact that its causation is multifactorial and that hormones cannot prevent all or even most fractures. Women are told some of the risk factors for osteoporosis, but these are so all-embracing (being female and being white are both commonly listed as risk factors) that most women come into the category of being 'at risk'.

More recently, cardiovascular disease has been appropriated as yet another serious indication for the use of hormones. Despite the fact that men are more at risk of this disease than women, cardiovascular disease is being represented as a result of oestrogen deficiency and, therefore, as another symptom of menopause. Consequently, women are being targeted to routinely use drugs to prevent cardiovascular disease. As with the osteoporosis marketing strategy, women are advised to use hormones long term to gain a beneficial effect on the heart. Any benefit shown in studies has been to current users of oestrogen alone (Stampfer et al., 1991). Consequently, doctors recommend that women use hormones for at least 30 years or even indefinitely. The advantage of this to the pharmaceutical industry is obvious.

There is still no clear evidence that oestrogen protects hearts. Despite the fact that there have been over 80 studies of this relationship published in English alone, the answer is not yet clear.

Indirect evidence points to a possible protective effect of oestrogen on hearts. Death rates for coronary heart disease are lower for women than for men. This has led to an assumption that a biological difference exists and that women are protected by endogenous oestrogens secreted from the ovaries. By this theory, the protective effect would be largely lost at menopause. This belief underlies the current support for the use of synthetic oestrogen to protect against heart disease.

On the other hand, sociological factors may explain the differences between men and women. There are two possibilities: lifestyle behaviours, and access to medical care (New England Research Institute (NERI), 1991). The fact that there is considerable variability across countries in the rate of heart disease by gender, with male to female ratios ranging from under two to over five, makes a sociological explanation plausible: 'Such variability is not consistent with a single biological cause' (NERI, 1991).

Some studies have shown a consistent gender difference in access to, and treatment by, the health care system (NERI, 1991). They found that even when people reported identical symptoms of chest pain and shortness of breath, men and women were treated differently. Women were less likely to have cardiovascular disease diagnosed or to have lifestyle changes, such as stopping smoking, recommended. Physicians were more likely to diagnose psychiatric disorders in women than in men. Women were therefore less likely to receive medical treatment.

The exact mechanism by which taking artificial oestrogens might protect hearts is not clear. Oestrogens might act directly on the artery walls, speeding the blood flow, and oestrogens are known to act favourably on blood lipids.

When it comes to studies showing an apparently beneficial effect of oestrogen on cardiovascular risk, it has been suggested that the difference may be explained by differences in women who use hormones and those who do not. According to a reviewer in the *Lancet*:

It is entirely possible that a woman who receives a prescription for hormone replacement is subtly healthier, or more determined to stay that way, than a woman who forgoes this therapy (Vanderbroucke, 1991).

Doctors might be less likely to prescribe HRT to women with existing hypertension or diabetes, or who are overweight, for instance. The positive results from studies could be explained by selection bias.

The manufacturers of hormones have been anxious to have protection against cardiovascular disease listed as an indication for the use of hormones. Following an application from Ayerst Laboratories, the Fertility and Maternal Health Drugs Advisory Committee of the FDA agreed in mid-1990 that the 'cardiovascular benefits of Premarin may outweigh the risks depending on the patient's risk profile for various estrogen-related disease and conditions' (Pearson, 1990). Ayerst is now seeking permission from the FDA to include prevention of coronary heart disease in women who have had hysterectomies as an indication for using Premarin. The reason Ayerst's request is confined to women with hysterectomies is that women who have a uterus must use combined therapy, with added progestogen, to protect the endometrium. It is not yet clear what this added progestogen will do to cardiovascular risk, but it is suspected that it might cancel out the beneficial effect of the oestrogen, or worse, add to the risk of cardiovascular disease. Progestogen is known to have an unfavourable effect on blood lipids.

This distinction is blurred by doctors who are now arguing for the universal use of hormones to protect hearts. It is also lost on women who do not understand the intricacies of the biological effects of hormones.

The conclusion of one consensus conference in 1988 was that 'the protective effect of estrogen on cardiovascular disease should be the major indication for [oestrogen] ERT' (Lobo and Whitehead, 1989).

Another widely quoted recommendation from a consensus conference was that of the Australian Menopause Society in 1991. The meeting concluded that:

As cardiovascular disorders rank as number one in the causes of major disability in later life, the consensus was that this was a compelling reason for considering long-term rather than short-term therapy (MacLennan, 1991).

Consensus conferences have become one of the major strategies for proselytizing about hormones. They have the air of impartial authority and are often widely reported in both the lay and medical media (Barnes, 1991; Dekker, 1991). The claim of 'consensus' is persuasive when recommendations are put forward, but it is often not clear what this actually means.

Consensus conferences are sometimes funded directly or indirectly by hormone manufacturers, and the invited participants, as workers in the menopause and osteoporosis fields, are weighted in favour of doctors who are positive about widespread hormone use. Their view of menopause is often distorted by the fact that their patient populations are not representative of women in the community.

The Australian Menopause Society is funded by the Australian Menopause Foundation, a body which includes representatives from both the medical profession and the drug industry. Dr Alastair MacLennan, who issued the consensus statement, is associate professor of obstetrics and gynaecology at the University of Adelaide and the producer of a video on HRT which was funded by the Upjohn Company and distributed in New Zealand by Ciba-Geigy (MacLennan, 1988).

Articles in the lay media are increasingly stressing the cardio-protective effects of HRT as a reason why women should use it (Dr Dan, 1991; Dekker, 1991). Women are promised longer life if they use hormones (Anon., 1991), despite the fact that this can only be claimed if an extremely optimistic view is taken of the breast cancer risk, and only for users of unopposed oestrogen.

The accumulation of evidence points to an added risk of breast cancer among users of menopausal hormones in the order of 30 per cent (Hulka, 1990; Hunt, Vessey, and McPherson, 1990; Steinberg et al., 1991). The risk is related to long-term use — of six to nine years and over — and to both opposed and unopposed oestrogen use. This period of use is the minimum recommended for prevention of bone loss. The new indications for hormone use — osteoporosis and protection against coronary heart disease — both require long-term, if not lifelong, use if the potential benefits promised by studies are to be approached. Women who succumb to arguments to use HRT as a strategy of disease prevention may be risking another disease which is a major killer of women in this age group, and a cause of disfigurement and psychological morbidity for those who have treatment.

Women are not necessarily aware of this. The risk of breast cancer tends to be glossed over in the patient information and media hype. Instead, women are urged to have regular mammograms, the implication being that this is all they need to do to protect themselves against cancer. Ironically, radiologists have said that mammography is less effective in women using menopausal hormones and have argued for more sparing use of HRT (Stomper, 1990).

Ayerst Laboratory's booklet for patients deals with the question of safety by simply referring women to their doctors. The word 'cancer' is not mentioned.

How safe is hormone replacement? Estrogen, like any medication, may cause side effects. (Please discuss the potential side effects with your physician) (Ayerst Laboratories, n.d.(b)).

Ciba-Geigy's patient booklet goes further. Although it does mention the word 'cancer', it says that whereas 'in the 1920s' high doses of oestrogen were found to increase the risk of certain cancers,

... present evidence indicates that a combination of low doses of the female hormones estrogen and progestogen can in fact have a protective effect in the occurrence of breast cancer, heart attacks and osteoporosis (Ciba-Geigy, n.d.).

Similar messages are to be found in handbooks for women, with the outcome that women are poorly informed about the real risks of hormone treatment. Neither are women given other information which may place the argument for hormone treatment in a broader perspective. For instance, women are often not told about the benefits of exercise in preventing bone loss, nor that oestrogen production does not stop at menopause, but in many women continues to take place by conversion of androgen into oestrogen in body fat.

Selling Hormones to Women

Promotion of HRT to women is mediated through the press, who are wooed by public relations firms representing drug manufacturers, and through physicians, who are subjected to a variety of marketing tactics. This twin-pronged attack primes women to consider and even ask for hormones, and encourages doctors to recommend hormones or at least be receptive to a request for treatment.

Articles for women about HRT and patient information literature invariably recommend further discussion with a doctor (Ayerst Laboratories, n.d.(b); Dr Dan, 1991; Donaldson, 1991). 'The first person you should see when you need help is your doctor', says a 'Patient Education Guide' on HRT published in *Patient Management* but supplied by Ciba-Geigy (Adis International Ltd, 1991). The drug companies are confident that doctors' recommendations will be consonant with their own commercial interests. An American study found that the physician's recommendation was the key factor in whether a woman was using or would use hormones (Ferguson, Hoegh, and Johnson, 1989). Ayerst Laboratories recognized the front-line value of doctors in recruiting users when it wrote that:

Ayerst believes that you, the medical practitioner, are the best equipped . . . to determine the need for therapy and the choice of therapy. And if that choice is estrogen replacement therapy; to outline the benefit/risk ratio. A public awareness program has been initiated urging women in their menopausal years to get the real story on the menopause, its cause and effect. And to get the facts from a professional source. You. The physician. Ayerst is confident that through proper counselling, patient education and estrogen replacement therapy, if necessary, you can help make life more liveable for patients in their menopausal years. And after (Ayerst Laboratories, 1991a).

Strategies for persuading doctors to prescribe hormones have included spear-heading postgraduate medical education on the menopause. For instance, in New Zealand, from the late 1980s Ciba-Geigy sponsored symposia on the management of menopause for general practitioners. These were conducted by reputable staff from the Family Planning Association and gynaecologists from National Women's Hospital menopause clinics. Doctors were also offered videos, both for themselves and for their patients. Similar tactics have been used by drug companies in other countries (Bungay, Vessey, and McPherson, 1980).

The tactic of using opinion leaders in the form of 'visiting experts' is another key strategy of drug companies. These people are hand-picked for their favourable view of the company's products and are toured around the country, at the drug company's expense, giving media interviews and addressing professional groups. The author's book on the marketing of HRT — *The Menopause Industry: a Guide to Medicine's 'Discovery' of the Mid-Life Woman* (Coney, 1991) was published on 4 September 1991. During September 1991, no less than three visiting medical experts toured New Zealand. Although all these tours were funded by the manufacturers of hormones, and media contacts were arranged by public relations firms, very few of the ensuing media reports mentioned how the visits were funded.

Conclusion

In all Western countries the use of menopausal hormones is on the rise. Thirty per cent of postmenopausal women in the US were using HRT by 1990, as were 10 per cent of British and Australian women (Roeber, 1990). In Britain, sales for HRT totalled £10 million by 1989 and had doubled over two years (Roberts, 1990). It was estimated that by the year 2000, a quarter of postmenopausal women in Britain would be using HRT, amounting to two million women. In Australia, the sales of oestrogens through retail pharmacies increased by nearly 52 per cent between 1981 and 1987 (Armstrong, 1988). In New Zealand, 77,000 prescriptions were written for HRT in 1989, up from 33,000 in 1983.

In the year ending March 1988, HRT cost the New Zealand Department of Health $1.45 million. By 1990, this had increased to $2.65 million and by 1991, to $3.66 million (Department of Health, 1991). Over 27,000 units of Estraderm transdermal oestrogen were sold to retail pharmacies in the year ending March 1990; by February 1991, this had leapt to 52,700, effectively doubling in a year. The percentage of menopausal women using hormones in New Zealand is variously put at 5–10 per cent.

Such dramatic increases in the use of hormones have been the result of a concerted campaign to increase the use of hormones among menopausal women. There have been three key factors in this:
- The activity of pharmaceutical companies in directly marketing HRT to doctors and in co-opting the field of medical education on the menopause.
- The enthusiastic promotion of HRT by 'champions' within the medical profession.
- The willingness of the lay media to uncritically mediate messages from the drug companies to women.

Women were vulnerable to this campaign to persuade them to use hormone replacement therapy because it exploited existing socially-caused fears about ageing and loss of status. Their prior relationship with the health system as well women facilitated acquiesence in the further medicalization of their lives. The information available to women from both medical and media sources was often biased, so that they were poorly placed to question or challenge the pressure to use hormones. The final chapter in the story of hormone replacement therapy will not be written until large-scale long-term independent studies prove or disprove the cancer risks.

References

Abbott Australasia Pty (1991) What nature takes away can now be brought back naturally (advertisement for Ogen) *Australian Medical Journal*.

Adis International Ltd (1990) Hormone Replacement Therapy Question Time (video companion booklet) Auckland.

Adis International (September 1991) Patient education guide: Hormone replacement therapy *Patient Management* 20 (9): 79–83.

Allen, W. (1974) Pros and cons of estrogen therapy for gynecologic conditions. In Reid, D. and Christian, C.D. (eds) *Controversy in Obstetrics and Gynecology* 785–93 Philadelphia, Saunders.

Anonymous (April 1976) Women and estrogens *FDA Consumer* 4–8.

Anonymous (2 July 1990) How to treat . . . menopause *New Zealand Doctor* 19–26.

Anonymous (15 January 1991) Medical frontiers: oestrogen helps to promote longer life *New Zealand Herald*.

Armstrong, B. (1988) Oestrogen therapy after the menopause — boon or bane? *Medical Journal of Australia* 148: 213–14.

Ayerst Laboratories (n.d.(a) *circa* 1986) Information for the patient (package insert) USA.

Ayerst Laboratories (n.d.(b) *circa* 1990) *Your guide to understanding menopause, estrogen deficiency and estrogen replacement therapy* (booklet) Auckland.

Ayerst Laboratories (1991a) She is woman (advertisement for Premarin) *Canadian Family Physician*.

Ayerst Laboratories (1991b) Women should have the chance to live the second half of their lives to the fullest (advertisement for Premarin) *Australian Medical Journal*.

Barlow, D.H., Brockie, J.A., Rees, C.M.P., and the Oxford General Practitioners Menopause Study Group (1991) Study of general practice consultations and menopausal problems *British Medical Journal* 302: 27-46.

Barnes, L. (7 October 1991) HRT suggested for all women *New Zealand Doctor* 1.

Bell, G. (30 July 1991) The selling of HRT *The Bulletin* 38-44.

Bell, S.E. (1987) Changing ideas: the medicalization of menopause *Social Science and Medicine* 24 (6): 535-42.

Benedek, T. (1950) Climacterium: a developmental stage *Psychosomatic Quarterly* 19: 1-27.

Bergkvist, L. et al. (1988) Risk factors for breast and endometrial cancer in a cohort of women treated with menopausal oestrogens *International Journal of Epidemiology* 17 (4):732-7.

Bonita, R. (1990) Hormone replacement therapy (unpublished) paper to Fertility Action Trust Politics of Women's Health course.

Boston Women's Health Book Collective (1984) *The New Our Bodies Ourselves* Ringwood, Victoria, Penguin Books.

Bungay, G.T., Vessey, M.P., and McPherson, C.K. (1980) Study of symptoms in middle life with special reference to the menopause *British Medical Journal* 281: 181-4.

Ciba (1990) *The Age of Confidence* (booklet) Auckland.

Ciba-Geigy (n.d.) *The Menopause — Knowing is Understanding* — (booklet) Auckland.

Coney, S. (1991) *The Menopause Industry: a Guide to Medicine's 'Discovery' of the Mid Life Woman* Auckland, Penguin.

Conrad, P. and Schneider, J.W. (1980) The medical control of deviance: Contests and consequences. In Roth, J. (ed) *Research in the Sociology of Health Care* Vol 1. Greenwich, Connecticut, JA1 Press.

Cooke, D. (1985) Social support and stressful life events during midlife *Maturitas* 7: 303-13.

Coope, J. and Roberts, D. (1990) A clinic for the prevention of osteoporosis in general practice *British Journal of General Practice* 40: 295-9.

Cooper, W. (1975) *No Change: A Biological Revolution for Women* London, Arrow Books.

Cooper, W. (January 1978) Combined therapy makes HRT safe as well as effective *Modern Medicine* 53-5.

Cooper, W. (1990) *Understanding Osteoporosis* London, Arrow Books.

Dejanikus, T. (May/June 1985) Major drug manufacturer funds osteoporosis public education campaign *Network News* 1-8.

Dekker, D. (11 September 1991) Soldiering on, or replacing the hormones *The Evening Post* 27.

Department of Health (November 1991) Costs of HRT to Department of Health, communication to author.

Deutsch, H. (1945) *The Psychology of Women* New York, Grune and Stratton.

Donaldson, A. (20 February 1991) Oestrogen . . . the menopause miracle *New Zealand Woman's Day* 28-9.

Dr Dan (9 October 1991) Your health: hot flushes *New Zealand Woman's Day*.

Draper, J. and Roland, M. (1990) Perimenopausal women's views on taking hormone replacement therapy to prevent osteoporosis *British Medical Journal* 300: 786-8.

Farquhar, C. (21 October 1991) The facts and fallacies of hormone replacement (review of *The Menopause Industry*) *New Zealand Doctor* 40.

Ferguson, K.J., Hoegh, C., and Johnson, S. (1989) *Archives of Internal Medicine* 149: 133-6.

Gorman, T. and Whitehead, M. (1989) *The Amarant Book of Hormone Replacement Therapy* London, Pan Books.

Greene, J. and Cooke, D. (1980) Life stress and symptoms at the climacterium *British Journal of Psychiatry* 136: 486-91.

Greer, G. (1991) *The Change: Women, Ageing and the Menopause* London, Hamish Hamilton.

Heine, W. (1991) The Pros and Cons of Hormone Replacement Therapy, address at the Auckland School of Medicine.

Hoover, R. et al. (1976) Menopausal estrogens and breast cancer *New England Journal of Medicine* 295: 401-5.

Howell, M. (1974) What medical schools teach about women *New England Journal of Medicine* 291 (6): 304-7.

Hulka, B. (1990) Hormone replacement therapy and the risk of breast cancer *CA — A Cancer Journal for Clinicians* 40 (5): 289-96.

Hunt, K. et al. (1987) Long-term surveillance of mortality and cancer incidence in women receiving hormone replacement therapy *British Journal of Obstetrics and Gynaecology* 94: 620-35.

Hunt, K. (1988) Perceived value of treatment among a group of long-term users of hormone replacement therapy *Journal of the Royal College of General Practitioners* 38: 398-401.

Hunt, K., Vessey, M., and McPherson, K. (1990) Mortality in a cohort of long-term users of hormone replacement therapy: an updated analysis *British Journal of Obstetrics and Gynaecology* 97: 1080-6.

Judd, H. and Utian, W. (1987) Introduction: what we hope to learn. Current perspectives in the management of the menopausal and postmenopausal patient *American Journal of Obstetrics and Gynecology* 156 (5): 1279-80.

Kadunce, D.P. et al. (1991) Cigarette smoking: risk factor for premature facial wrinkling *Annals of Internal Medicine* 114: 840-4.

Kaufert, P.A. (1980) The perimenopausal woman and her use of health services *Maturitas* 2: 191-205.

Kaufert, P.A. (1982) Myth and menopause *Sociology of Health and Illness* 4 (2): 141-66.

Kaufert, P.A. and McKinlay, S. (1985) Estrogen replacement therapy: the production of medical knowledge and the emergence of policy. In Lewin, E. and Olesen, V. (eds) *Women, Health and Healing* 113-38 London, Tavistock Press.

Kaufert, P.A. and Gilbert, P. (1986) Women, menopause and medicalization *Culture, Medicine and Psychiatry* 10: 7-21.

Law, M.R., Wald, N.J., and Meade, T.W. (1991) Strategies for prevention of osteoporosis and hip fracture *British Medical Journal* 303: 453-9.

Lawrence, C. (1991) Mother Premarin. In Taylor, D. and Coverdale Sumrall, A. (eds) *Women of the 14th Moon: writings on Menopause* 51-3 California, The Crossing Press.

Lennane, K.J. and Lennane, R.J. (1973) Alleged psychogenic disorders in women: a possible manifestation of sexual prejudice *New England Journal of Medicine* 288: 288-92.

Lobo, R.A. and Whitehead, M. (1989) Too much of a good thing? Use of Progestogens in the menopause: an international consensus statement *Fertility and Sterility* 51 (2): 229-31.

McCrea, F. (1983) The politics of menopause: the 'discovery' of a deficiency disease *Social Problems* 31 (1): 111-23.

McKenzie, N. (28 February 1990) Interview with Wayne Mowat, National Radio.

McKinlay, J., McKinlay, S., and Brambilla, D. (1987) The relative contributions of endocrine changes and social circumstances to depression in mid-aged women *Journal of Health and Social Behaviour* 28 (4): 345-63.

McKintyre, S. and Oldman, D. (1977) Coping with migraine. In Davis, A. and Horobin, G. *Medical Encounters* 55-71 London, Croom Helm.

MacLennan, A. and The Queen Victoria Hospital Foundation Inc. (1988) Understanding the menopause and hormone replacement therapy (video) funded by the Upjohn Company, distributed in New Zealand by Ciba-Geigy.

MacLennan, A.H. (1991) Consensus statement: Hormone replacement therapy and the menopause *Medical Journal of Australia* 155: 43-4.

Mishell, D.R. (1989) Is routine use of estrogen indicated in postmenopausal women? *Journal of Family Practice* 29 (4) 406-15.

Nachtigall, L. and Heilman, J. (1986) *Estrogen: The Facts Can Change Your Life* Los Angeles, The Body Press.

New England Research Institute (NERI) (Summer 1991) Gender differences in heart disease: biological or social? *Network News* 1-3.

Notelovitz, M. and Ware, M. (1985) *Stand Tall* New York, Bantam Books.

Pearson, C. (July/Aug 1990) FDA waffles on Premarin decision *Network News* 1-7.

Persson, I. et al. (1983) Practice and patterns of estrogen treatment in climacteric women in a Swedish population *Acta Obstetrics and Gynecology of Scandinavia* 62: 289-96.

Polivy, J. (1974) Psychological reactions to hysterectomy: a critical review *American Journal of Obstetrics and Gynaecology* 118 (3): 417-26.

Pollner, F. (14 January 1985) Osteoporosis: looking at the whole picture *Medical World News* 38-58.

Rauramo, L. (1986) A review of study findings of the risks and benefits of oestrogen therapy in the female climacteric *Maturitas* 8: 177-87.

Report from the Boston Collaborative Drug Surveillance Program (1974) Surgically confirmed gallbladder disease, venous thromboembolism, and breast tumours in relation to postmenopausal estrogen therapy *New England Journal of Medicine* 290 (1): 15-19.

Roberts, Y. (12 January 1990) Fortysomething *New Statesman and Society*: 1214.

Roeber, J. (June 1990) The forever therapy? *Vogue* 17-23.

Schering (1975) For more years of femininity and joie de vivre (advertisement for Progynova) *New Zealand Medical Journal*.

Schering (1990) So a woman can enjoy being a woman (advertisement for Progynova) *New Zealand Doctor*.

Scully, D. and Bart, P. (1973) A funny thing happened on the way to the orifice: women in gynaecology textbooks *American Journal of Sociology* 78 (4): 1045-50.

Seaman, B. and Seaman, G. (1977) *Women and the Crisis in Sex Hormones* New York, Bantam Books.

Silverman, M. and Lee, P. (1974) *Pills, Profits and Politics* Berkeley, University of California Press.

Smith, D.C. et al. (1975) Association of exogenous estrogen and endometrial carcinoma *New England Journal of Medicine* 293: 1164-7.

Sontag. S. (February 1974) Women grow old but men mature *Broadsheet* 16: 5-6.

Stampfer, M.J. et al. (1991) Postmenopausal estrogen therapy in cardiovascular disease: 10-year followup from the Nurses' Health Study *New England Journal of Medicine* 325: 756-62.

Steinberg, K.K. et al. (1991) A meta analysis of the effect of estrogen replacement therapy on the risk of breast cancer *Journal of the American Medical Association* 265: 1985-90.

Stomper, P. et al. (1990) Mammographic changes associated with postmenopausal hormone replacement therapy: a longitudinal study *Radiology* 175: 487-90.

Utian, W. (1987) The fate of the untreated menopause *Obstetrics and Gynaecology Clinics of North America* 14 (1): 1-12.

Vanderbroucke, J.P. (1991) Postmenopausal oestrogens and cardioprotection *Lancet* 337: 833-4.

Van Keep, P. (1990) The history and rationale of hormone replacement therapy *Maturitas* 12: 163-70.

Wilson, R.A. and Wilson, T.A. (1963) The fate of the non-treated postmenopausal woman: a plea for the maintenance of adequate estrogen from puberty to grave *Journal of the American Geriatrics Society* 11: 347-62.

Wilson, R.A. (1966) *Forever Feminine* London, Allen.

Wolfe, N. (1990) *The Beauty Myth* London, Chatto and Windus.

Wren, B. (March 1987) Oestrogen therapy after menopause: a viewpoint on its rational use *Current Therapeutics* 25-36.

Youngs, D. (1990) Some misconceptions concerning the menopause *Obstetrics and Gynaecology Clinics of North America* 75 (5): 881-3.

Ziel, H. and Finkle, W. (1975) Increased risk of endometrial carcinoma among users of conjugated estrogens *New England Journal of Medicine* 293: 1167-70.

Choosing a Remedy

Alex Thomson

The pharmaceutical industry is a major figure in the health sector of western nations. In New Zealand, as elsewhere, there are two fairly distinct market segments in the industry's activities — over-the-counter (non-prescription) items, and prescription-only medications. Traditionally the former are marketed to the public, the latter to the medical profession. However, there have been recent examples of the industry advertising prescription-only medications in public news media. These advertisements are designed to increase public demand for pharmaceutical treatments, often in cases where medical indications are rare, for example the treatment of male pattern baldness. Such advertising is at present a small component of the advertising of prescription-only medications.

The industry's marketing activities in the over-the-counter and prescription sectors have different implications for New Zealand's health service. Over-the-counter medications are a direct cost to the consumer, but do not directly add to the health bill as it affects insurers or the State. The marketing of over-the-counter medications may, however, reinforce the notion that all ills have a pharmacologic solution, and thus increase public demand for prescription medications. Prescribed medications, on the other hand, are a direct cost to the State in almost all cases, with a small part of the cost borne by consumers and, in some cases, health insurers.

The relationship between the pharmaceutical industry and the consumer, in the case of over-the-counter medications, may be considered to be a normal commercial transaction where the principle of *caveat emptor* may apply. However, in the case of prescription medication there is a third party between consumer and industry — a professional who is trusted by the consumer (and the State) to provide scientifically sound information about the most appropriate remedy, who controls access to the field of prescription medications, and who determines in large part which product within this field will be used. How does this professional, the medical practitioner, choose the remedy? Is there evidence that this choice is made on the basis of sound scientific evidence, or are other influences at work?

This chapter will explore factors in the education and socialization of medical practitioners which modify their choice of therapeutic remedies. Over-the-counter pharmaceutical sales will not be discussed further. To clarify the picture the chapter will focus on the general (or family practitioner) as the individual responsible for advising patients/consumers about the majority of their health care needs.

The Socialization of the General Practitioner

The relationship between the pharmaceutical industry and practitioners may start soon after entry to medical school and continues throughout their practising lifetimes (see Figure). Through industry courting of the profession, and through the profession's acceptance of courting by the industry, the industry gains a significant place in the social milieu of the medical practitioner.

The breadth of Pharmaceutical Industry Funding for the Medical Profession for Promotional Purposes

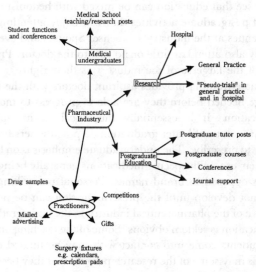

Medical education can be considered in two phases — basic or undergraduate training, and the subsequent postgraduate and continuing professional education. The pharmaceutical industry plays a significant role in the lifelong education of the medical practitioner. With activities as diverse as funding teaching posts and social events, sponsoring research and conferences, and giving doctors free drug samples and bottles of wine, they are a part of the formal education and socialization of the medical profession.

From the time of entry to medical school students solicit, and are solicited by, the pharmaceutical industry. The industry recognizes that medical students will bloom into the important controllers of their products. Sponsorship of activities such as medical student balls and publications by the industry is not seen to be matched in other university faculties where the future graduates will not control access to pharmaceutical expenditure. However, this is not to argue that if the industry spread its activities equally across all faculties, it would be ethical for medical students to accept such funding. As will be seen from later arguments it is only in the non-medical faculties that pharmaceutical industry funding may be ethically and morally acceptable. Nevertheless, medical school staff at least tacitly approve of the industry's sponsorship of student activities. The staff act as role models for the students and certainly do not condemn soliciting for or accepting donations. In fact, the students can watch the senior members of the profession pursuing the same funding source with some skill. As students graduate they see that education can be mixed with hedonism. Through sponsorship of postgraduate activities they can enjoy attending meetings in exclusive venues at the industry's expense. Spouses and even children may frequently also attend at little or no cost to the doctor. Practitioners come to accept the largesse of the industry as their right.

Undergraduate training provides aspirant doctors with the basic skills and knowledge needed before they are able to progress to more focused specialty education. It is essentially theoretical in its approach to prescribing, since not until after graduation do practitioners first gain the legal authority to prescribe. The undergraduate emphasis is on broad drug groups and their use, with specific medications generally being discussed by generic, as opposed to brand names. Personal prescribing patterns, however, cannot develop until the prescribing skill can be practised.

The influence of the pharmaceutical industry in the undergraduate phase of medical education is seldom obvious. Some of the teaching and research staff whom students come into contact with may be funded directly by the industry, as may some of the research projects that they become aware of. This type of sponsorship is not directly visible since the source of funds for staff and research is generally not acknowledged during teaching activities. There is no evidence that such funding influences the education process in any overt way.

In the initial period after graduation, practitioners' prescribing is strongly influenced by their working environment, that of the hospital service. The nature of the patients seen and treated in the hospital tends to favour the prescription of newer or more powerful medications, since patients who are well controlled or treated in the community with older or less powerful medications are less likely to be hospitalized. Secondly, the hospital service is the area in which most new medications undergo trials. Thus these medications are often being used even where older medications would be acceptable substitutes. Thirdly, the prescribing of the specialists is influenced by the same pharmaceutical industry marketing ploys as is that of the general practitioners. In addition, specialists frequently receive industry funding for trips to overseas conferences, in which the industry may buy whole sessions designed to promote their products.

These early prescribing influences will affect the attitude of doctors entering general practice, and will tend to steer them towards newer rather than older medications. Doctors within general practice are indirectly subject to the same influences. As patients are discharged from hospital the general practitioner is expected to continue the medications which have been prescribed by the hospital doctors. The prestige of specialist and hospital means that these new medications are thought of as being more effective — this influences the general practitioner to choose them for the next patient with a similar condition.

About two-thirds of doctors who elect to enter general practice decide to undertake the general practice training scheme, a one-year formal education programme. This programme, a mixture of seminars and apprenticeship-type learning during which the trainees are attached to experienced practitioners, is largely funded by a grant from the Government. Sponsorship from the pharmaceutical industry is received for some activities. The implications of this sponsorship will be discussed later. Formal training for general practice is not mandatory in New Zealand, in contrast to the situation in many areas of the world. Nevertheless, the medical profession and the Government consider such training highly desirable. Unfortunately, the New Zealand Government has significantly reduced its funding support for the training programme, a move in line with its philosophy that the cost of education should be borne by the consumer. This shift in policy may see an increased demand for pharmaceutical sponsorship in the programme with the attendant ill effects of such sponsorship.

For all other practitioners, and for those leaving the training programme, continuing professional education receives virtually no Government funding. There are many modes of continuing professional education that are utilized by practitioners. Practitioners state that they prefer reading journals and attending lectures and seminars (Thomson, 1990a; Barham and Benseman, 1981) but it does not necessarily follow that modes of continuing professional education that receive 'most favoured status' are the most effective delivery modes in terms of their ability to change knowledge, attitudes or behaviour (Lloyd and Abrahamson, 1979).

Commercially-sponsored review journals provide general practitioners with an important source of written information on health care issues. These sources are rated highly since practitioners, especially those in their formative early years, see the presented information as more relevant than that presented by the 'scientific' journals. In contrast, however, the experienced general practitioners who teach in the general practice training programme rate scientific journals as highly as they rate the review journals (Thomson, 1990a). The commercial review journals are free to practitioners, and the costs of publication are paid for by the advertisers, which are all pharmaceutical companies. In order to attract advertisers, the publishers choose topics that cover areas in which a company may have interests, and sell advertising space around the articles. While their actual content may not be influenced by the pharmaceutical companies, the articles tend to deal with issues or aspects of issues that require a pharmacological approach. The publishers believe that this is necessary if they are to successfully attract the advertising dollars that they depend on. A bias in subject matter, if not in the content within subjects, is clearly apparent.

Medical meetings are the other education method most favoured by practitioners. Such meetings may be organized and funded by a variety of methods, but Government funding is almost never available. Meetings are usually held in evenings or weekends to prevent clashes with clinical commitments. Practitioners therefore often look to these meetings as a chance to mix social contact with education since they offer attractive venues, food, and drink. These additions are costly and pharmaceutical sponsorship is often used to reduce the cost to participants. Some meetings are organized entirely by companies which choose the topics to be discussed, the speakers, the venue and the various 'sweeteners'. These meetings are designed around the companies' needs and interests, not the needs of practitioners or patients, but are nevertheless popular with practitioners. More frequently a lower level of sponsorship occurs when the rights to be associated with topic areas are sold to a company, but the content within the area is organized by the practitioners themselves, or by continuing professional education organizations. Such sponsorship is permitted under the recommended guidelines for meetings which involve pharmaceutical companies (New Zealand Council for Postgraduate Medical Education, 1983). Many of the medical educators involved in continuing professional education courses or conferences have expressed concern that meetings that receive no sponsorship may be less popular with practitioners because the organizers cannot offer associated 'sweeteners' without charging participants.

The expectations of practitioners and the cost of inducements make pharmaceutical sponsorship an important factor in designing conferences and courses, and, for at least some practitioners, determining their attendance at meetings. The effects of the sponsorship will be explored later.

A significant source of education for practitioners, the importance of which is seldom recognized by participants, is the direct contact between pharmaceutical representatives and practitioners. Most companies try to visit each practitioner a number of times each year to raise the profile of their products. These visits are designed to increase prescribing by several methods. Firstly, the visit provides a reminder of company and product names that leave the names well-placed in the doctor's subconscious. Secondly, at these visits they provide information designed to increase the usage of their products over that of competing products. There is no reason to believe that inaccurate information is given to practitioners, but the information may be clearly biased to demonstrate the good features, while ignoring the less than ideal features of their products (Holmes, 1977). There is good evidence that this approach is successful in altering prescribing behaviour (Avorn, Chen, and Hartley, 1982; New Zealand Board of Health, 1973). In addition, companies continually bombard doctors with written information (Durham, 1990) which reinforces their name and stresses single points that may make their products seem attractive. Personal visiting and written information are accompanied by sweeteners such as free give-aways: from items as simple as pad or pen, to expensive textbooks, wine, and chances to win overseas holidays or mobile telephones (St George, 1989). Doctors freely admit that these inducements do play a part in their readiness to accept visits or to take part in other advertising ploys, and may be seen by them as a fair reward for the time they spend with the pharmaceutical representative or in reading the literature. In accepting even small gifts practitioners may not recognize the price that they, and the State, are paying. The ubiquitous placement of product names in front of the doctor on pens and pads is not noted on a conscious level but is likely to subtly change prescribing behaviour.

Free samples of medications to trial are frequently given to doctors by the representatives. Such trials can never have scientific validity. They have a direct educational or marketing function by providing further impetus for prescribing change. A doctor's prescribing behaviour tends to be affected by the results of treatment of the last patient that they saw. In fact, probably only if the outcome was adverse, or the condition chronic, would the patient be likely to re-attend. Given that the results of the trial treatment are likely to be little different from the treatments formerly used by the practitioner, the patient is unlikely to re-attend or offer new insights. There is little utility for the practitioner in such a trial of therapy. When patients are taking well-tolerated ongoing medications there is no clinical indication for change; thus, in exchange for a company-funded trial of a few weeks or a few months duration, the patient will normally be continued with the medication but then at a cost to the Government and the consumer. The high cost of pharmaceutical promotion (Kawachi, 1990) must be recovered from sales, either by increases in the cost of each individual pharmaceutical or by the industry influencing doctors to increase their prescribing. The result may be seen in the inflation of the pharmaceutical expenditure in New Zealand in the last 10 years (see Table below, data from the New Zealand Department of Health, 1991).

Table
New Zealand Pharmaceutical Expenditure (1983–1990)

	1983	1984	1985	1986	1987	1988	1989	1990	Period 1983–90 (cumulative)
Total expenditure ($ \times 10^6$)	196.1	220.6	254.8	354.3	448.6	516.7	559.0	572.3	
Cost to Government ($ \times 10^6$)	196.1	220.6	254.8	346.3	439.6	506.7	549.0	517.3	
Annual rate of increase in total expenditure (%)	—	12.5	15.5	39.0	26.6	15.3	8.4	2.4	191.8
Annual CPI increase (%)	—	9.4	15.3	18.2	9.6	4.7	7.2	4.9	92.5
Annual population increase (%)	—	1.0	0.7	0.2	0.9	0.8	0.3	0.9	5.0

The information presented in the pharmaceutical advertising-supported journals and by pharmaceutical representatives is the most potent source of influence for new prescription of pharmaceutical agents, with the messages being constantly reinforced by the brand-name-tagged give-aways that clutter the doctor's office and consulting room. Fogel (1989) demonstrated that almost 750 promotional items were found in each consulting room in a Canadian training family practice. Other sources of continuing professional education tend not to deal with prescribing issues in depth since these are well covered by the pharmaceutical industry, and frequently with such attractive 'sweeteners' that the non-industry educators cannot hope to compete. Education sessions about issues that do not involve pharmaceutical considerations, such as ethnic issues, seldom receive the attention or sponsorship of the pharmaceutical industry.

Key Issues

When a patient requires treatment the medical practitioner ideally should answer two questions before deciding on the most appropriate form of treatment:

(1) What is the cost to the State of each treatment option and its effect on overall health expenditure?

(2) What are the likely outcomes for the patient of each treatment option?

The doctor can easily calculate the cost of treatment but seldom bothers to, since the difference between a prescription item costing, for example $15.80 and $18.80, seems trivial in the context of the individual patient, especially since the patient does not directly bear the cost. The recent increase in tax payable on prescriptions means that there is now a positive financial incentive to prescribe items that have a retail cost less than the tax on each item. If, however, the choice is between items which are more expensive than the amount of tax there is no incentive to prescribe the cheaper medication. The effect of the tax on prescribing costs is at present uncertain. There is as yet no evidence that the tax will produce a trend to prescribe lower-cost medications across the whole spectrum of costs. The effect on the State's total health expenditure of an individual doctor's decision to prescribe a more expensive medication is perhaps clear, but may not be considered by the prescribing doctors. There is no evidence available to them which clearly links more expensive prescribing with a decrease in the availability of other needed health care. In addition, prescribing doctors have their focus on the immediate needs of the patients in front of them, rather than on more global issues.

The second question about the outcome of treatment should, on the face of it, be easily answered, but data are seldom available. There may be evidence that drug A causes fewer side-effects than drug B, or that drug A covers a wider spectrum of bacteria than drug B. Fewer side-effects may, however, be practically unimportant. New drugs may seem to have fewer side-effects because there has been less time for side-effects to be reported. A wide-spectrum drug may not be needed. In an attempt to do the best for the patient, the doctor will tend to prescribe drug A if it seems theoretically better. Data on the outcome of alternative treatments often show that there is no practical difference between medications. When there are many similar products the doctor will tend to prescribe the one that comes readily to mind. Both the recall of side-effects and recall of name are heavily influenced by the promotional activity of the pharmaceutical companies, particularly as their information is not balanced by any other source of comparable information.

Medical Research

The answer to practitioners' questions about which is the most appropriate therapy may be unavailable for another fundamental reason, involving the funding of medical research. As with medical treatment, medical research is subject to financial, resource, and ethical constraints (Thomson, 1990b) which limit research activity. There is a trend in New Zealand and other world economies to a user-pays, market-driven economy. Research funds have not kept pace with need and are increasingly expected to be found from 'the market'. Even within the universities the user-pays philosophy has become pervasive. The major source of 'market funds' in health care research is the pharmaceutical industry. However, their interest is, by the nature of their business, narrow. The research they are prepared to fund is subject to a number of constraining factors:

(1) Pharmaceutical research usually compares one product with another, or with a placebo. These comparisons are dictated by commercial need, or by the need to comply with the Federal Drug Administration or the New Zealand Health Department guidelines regarding clinical trials. However, within the options for the treatment of a particular condition, there are frequently non-pharmacologic options — as a part of treatment or as the complete therapy. There is no advantage, and often some disadvantage, for a pharmaceutical company in adding a study arm that contains non-drug treatment. An additional treatment arm complicates the study's design, increases study costs, demands increased staffing and patient resources, and may produce findings that are not in the industry's interests. Placebo treatment is not equivalent to non-pharmaceutical therapy. Placebos have been shown to be potent pharmacologic agents demonstrating the power of the patient's expectation of a result from therapy.

(2) Pharmaceutical research concentrates on conditions in which new drug treatments are or may be available. Many unanswered questions may exist in other areas where no new treatment is available for trial, such as in the optimal use of old medications (for example, many 'older' four times a day medications may be just as effective in twice daily dosage as some of the newer long-acting or slow-release preparations).

(3) Pharmaceutical research concentrates on conditions in which a pharmaceutical treatment exists. Many key issues in medicine do not have drug solutions, such as conditions in which behaviour modification is the appropriate management (for example, the management of stress).

(4) Pharmaceutical research concentrates on the aspects of a condition that are amenable to drug therapy. Other aspects, such as the problems related to diagnosis, are not likely to be areas of interest for the industry.

(5) Pharmaceutical research uses resources that cannot easily be replaced. In particular, patients are a finite resource in terms of their numbers, available time, and willingness to be involved. General practitioners receive many requests for their involvement in research from a variety of other sources in addition to the pharmaceutical industry. Each request takes some of the general practitioner's time, so practitioners can be involved in a limited number of studies. Research in one area reduces the potential for research in another area. In addition, the New Zealand Health Department may insist on local pharmaceutical trials, or proof of need, before licensing a product, even if such studies have been satisfactorily carried out in another environment; there is pressure, therefore, to carry out unnecessary trials to meet licensing requirements. The industry gains an additional benefit, as these trials give potential marketing advantages. The pharmaceutical industry has at times used payments or inducements which are significantly higher than can be justified on the basis of the time that the general practitioner may spend on the study. Such inducements are justified by the need to encourage general practitioners to spend their time on the study, and to pay them for the inconvenience and their expertise. However, these inducements may tend to distort the cost structure of research in general practice — further compounding the problem of low resources for non-market-oriented research.

(6) Pharmaceutical-funded research may produce commercially sensitive information. Restrictions may be placed upon data ownership and analysis, and on publication. Such restrictions may not be in the best interests of either the research community or the public.

(7) Pharmaceutical research seldom looks at major new advances in therapy. Medical knowledge generally increases in small, although often

significant, steps. Many pharmaceutical trials involve the testing of products that are only marginally different from others currently marketed.

(8) Despite much pharmaceutical research yielding little advance in knowledge it may consume significant resources. To demonstrate the efficacy of a new product requires large numbers of trial subjects and careful attention to measurement of outcome. Especially in the early phases of such research close monitoring of the subjects is required in order to detect unexpected adverse events. Most pharmaceutical studies are thus carried out in the hospital setting. The results of treatment in this setting may not be applicable to the general practice setting, although this is not recognized by many specialist or pharmaceutical company educators. General practitioners generally see illnesses in an earlier stage than do the hospital doctors or other specialists. It is not possible, for example, to extrapolate from the management of depression in a psychiatric hospital or clinic to the treatment of depression in general practice; in general practice the illness is likely to be milder and the rates of spontaneous resolution will be higher.

With such constraints, the results of pharmaceutical-sponsored research often leave fundamental questions unanswered. Unfortunately, such questions are infrequently the subject of other funded research since they deal with areas in which a 'satisfactory' (normally pharmaceutical) approach to management has been established. Practitioners are then left without adequate information on which to answer their key questions! There is thus no balance or foil for the 'first thing that comes to mind' approach to prescribing. It seems that the major influences on prescribing are educative rather than cost factors — but with a major rationale behind much of the educative effort being a desire to enhance brand name awareness, there are serious difficulties in the delivery and effects of this education.

The Commercial Environment

In order to determine what steps may be able to be taken to remedy the existing situation we need to explore the underlying legal, ethical and commercial/fiscal responsibilities of the pharmaceutical industry and the medical profession. The pharmaceutical industry is a commercial enterprise required to return an adequate dividend for shareholders on their companies' invested funds. To achieve this requires success in their chosen area of commercial endeavour, the health care arena. They therefore carry out a range of activities which include some or all of the following:

(1) Developing new products that meet clearly defined health needs in areas which the public and Government accord some priority.

(2) Marketing new products before patent life expires so that the maximum market advantage can be realized from the product before competition reduces their market share or the price.

(3) Keeping the profile and image of their products high so that these will continue to be prescribed by doctors. If products prove to have dangerous side-effects, and a company does not handle the situation with integrity, then it stands to lose money directly through legal costs and indirectly through a damaged reputation. Thus there is commercial pressure to produce safe, effective products and market them wisely.

Legal responsibilities also exist: companies must obtain approval to register and market products in any country, and they must keep adequate records of reported side-effects. To meet these legal responsibilities the companies must invest in research and support for their products. They are also subject to ethical guidelines in their dealing with the public and the profession. These guidelines are set out by organizations such as the Researched Medicines Industry in New Zealand (Pharmaceutical Manufacturers Association of New Zealand Inc., 1984), but are not legally binding. Ethical boundaries also appear to shift as they are tested by the various companies, and what was acceptable 10 years ago may not be acceptable now. For example, while informed consent guidelines have become stricter, there has been an increasing acceptance of gifts to doctors in exchange for the right of access to a little of their time. Adherence to the ethical guidelines is probably maintained as much by commercial reality as by an underlying ethical belief system; a company that appeared to be stretching the ethical boundaries beyond the profession's or public's tolerance would be in danger of harming its reputation and sales.

It is clear that contact between the pharmaceutical industry's sales staff/representatives and the practitioner influences prescribing in favour of that company's products. There is no evidence that the profession is encouraged to, or does, prescribe without some reason to do so (McLauchlan, 1990). However, the pharmaceutical industry has a vested interest in a pharmaceutical solution (Thomson, 1990c; Thomson, 1990d; Hemminki, 1977). Their formal education and research efforts are generally of a high quality, but they reflect this industry bias. The industry is not being unethical. Nevertheless, to the extent that the association with the industry influences the profession's prescribing, the profession begins practising medicine with an unwitting bias. The profession must therefore be careful to maintain as separate an existence from the industry as possible. The existence of each, however, is dependent on the other. The relationship therefore must be conducted within careful ethical guidelines. The industry will continue to follow the commercial ethic. Its shareholders will demand this. Our concern must be with the ethical behaviour of the professionals.

For medical practitioners there is little commercial constraint on prescribing. The cost of medication has no obvious direct effect on the practitioner's income — there is no positive or negative benefit to the doctor in constraining or inflating prescribing costs. Even with the new prescription tax it is far from certain that market forces will drive patients to doctors who prescribe either more cheaply (in the small sector of the prescription market where the patients will be affected), or less frequently, and thus force doctors to review their prescribing. Patients swap 'doctor experiences' and thus freely share information about doctor behaviour and costs, although recent experience has demonstrated that cost factors alone do not alter patient choice of doctor to a significant extent. Practitioners' fiscal responsibility for the effects of their prescribing on overall health costs has not been defined by the State. However, Ministers of Health and the Health Department frequently comment on the need for cost-conscious, responsible prescribing, and bemoan the escalating pharmaceutical bill.

As can be seen from the Table on page 214, pharmaceutical expenditure has increased much faster than the underlying rate of inflation. Most of the escalating pharmaceutical bill has been met by Government, and the remainder by patients, while a proportion of the latter will be reimbursed by insurance companies. As with all other areas of health expenditure there would seem to be considerable room for saving, with much international (McLauchlan, 1991) and inter-doctor variability in the cost of prescription medication without a corresponding effect on the standard of health. Health care of those not in hospital, including those attending general practitioners and clinic services, is responsible for approximately 88 per cent of the total pharmaceutical bill in New Zealand (New Zealand Department of Health, Health Benefits Section, 1991), so the prescribing habits of general practitioners would seem to be the area of greatest potential in reducing the pharmaceutical bill. However, as already noted, many of the most expensive long-term medications that patients take are initially started in hospitals. Thus it may not be possible to dramatically reduce general practice prescribing costs or to rationalize such prescribing simply by focusing on the general practitioner. There has been feedback to doctors about the average cost of their prescriptions with a measure of peer comparison, but until this can be linked to diagnosis, patient age breakdown, and the level of non-prescribing by doctors, the feedback will generally be ignored as irrelevant.

Legal and Ethical Considerations

Legal considerations do not influence prescribing in any major way. The regulations governing prescribing (New Zealand Drug Tariff, 1990) put limits on the quantity of medications that can be prescribed at any time, which vary between different classes of medication. These regulations limit neither the patient's possible access to medication nor the practitioner's overall prescribing potential. There is a legal obligation on the practitioner to provide adequate care, as defined by case law. Two principles stand out:
(1) The patient must be provided with adequate information about the risks and benefits of treatment.
(2) The care is judged in relation to the standard provided by peers.

As we have already seen, there is a paucity of adequate information on which to assess the risks and benefits of treatment. Comparison with peers requires adequate standards for the profession as a whole. If we accept that the profession is being adversely influenced by the pharmaceutical industry it could be argued that the lack of an absolute standard of care and behaviour is unfortunate. Constructing such a standard in a fashion that could be effectively administered would pose major difficulty.

Without a system of positive or negative incentives to control prescribing, and in the absence of effective legal constraints on prescribing, we must consider the potential effectiveness of ethical considerations to influence prescribing.

It has been argued that the fundamental ethic of the pharmaceutical industry is a commercial one, tempered as necessary to maintain a reputation that allows them to continue to function in the health care industry. It has also been seen that educative influences are important in determining prescribing choice, and that the pharmaceutical industry has a major input into the sponsoring and running of education programmes, and the research on which new information about treatments is based. Given the pivotal position of the pharmaceutical industry, how should the medical profession respond to the ethics of their relationship with the industry?

Generally the ethical code of the medical profession is silent about this relationship. However, the New Zealand Medical Association's Ethical Code contains the statement that 'Motives of profit shall never be permitted to influence the free and independent exercise of professional judgement on behalf of a patient' (New Zealand Medical Association, 1990). To the extent that the association with the industry profits the profession, and influences the profession's judgement or behaviour, it must be unethical. This has not been tested, however, since the profession judges its members against the behaviour of their peers. Where the association does not lead to profit, but behaviour is influenced, the Code is silent. The New Zealand Council for Postgraduate Medical Education (1983) developed clear guidelines for the relationship between the parties as it relates to education. These guidelines cover several areas:

(1) Speakers and subjects should be chosen by the continuing professional education organizations on the basis of educational needs.

(2) All arrangements for courses should be undertaken by the continuing professional education organization.

However:

(3) Sponsorship by associating the name of, for example, a pharmaceutical company with a topic or speaker, or advertising in the back of the programme is acceptable.

These guidelines are frequently breached both within and without the Medical Schools, and it seems that many medical educators remain ignorant of the guidelines (Murdoch, 1990). There may have been inadequate promulgation of the guidelines which, at best, are an uneasy compromise. To adapt a quip of George Bernard Shaw, they do not alter the fundamental prostitution of the medical profession, but merely its price! In addition, the guidelines have not kept pace with recent trends. Talk-back radio shows with a medical flavour are a recent phenomenon. The Medical Association and a number of individual practitioners have been associated with these. On some shows the radio station has sold time to advertisers from the pharmaceutical industry, allowing the theme of the day to be associated with one company's solution to the problem. This is likely to be seen by the public as an endorsement by the profession of the company's product. Direct product endorsement, or appearance in advertising for any product, pharmaceutical or otherwise, is proscribed by the profession's Ethical Code, but this, however, appears to fall outside the Code. It does not directly involve the professional in advertising the product; nor does it fit within the intent of the New Zealand Council for Postgraduate Medical Education Code which is concerned with doctor education, not public education.

Public education material that is distributed in written form is often sponsored by pharmaceutical companies. In many cases there appears to be a bias in the information presented; that is, the information is not inaccurate, merely slanted or incomplete. The provision by a professional of such information, whether biased or not, together with a company's advertising, may lend additional credibility to the product advertised. For some time biased information on infant feeding and bottle sterilization has routinely been given to mothers on discharge from many of New Zealand's obstetric hospitals. Such advice, while not emanating from the doctor, is imbued with authoritative credibility because of the setting in which it is dispensed. There is a clear need for a new ethical code to cover the area of public education.

To produce a change in the behaviour of the medical profession will not be easy. From their entry to medical school, students and doctors court, and are courted by the pharmaceutical industry. Doctors generally deny that such a relationship may affect their judgement or be ethically questionable. Those running conferences soon find that attendances are reduced unless the venue is attractive, the meals lavish, time off ample, and spouses are invited. Some educators involved in running night or weekend courses express similar concerns. The true cost of education seems to go up, while the doctors' desire to pay goes down. Sponsorship is necessary if one is to remain in the educational business. However, New Zealand's biggest continuing professional education organization for general practice (the Goodfellow Unit) only accepts sponsorship for the administrative costs involved in running courses. Despite their not using expensive venues and providing free meals, course attendances are high (Barham, 1991). Thus course participants pay the majority of the cost of their education. This is heartening, suggesting that it may be possible to divorce the pharmaceutical industry from the continuing professional education organizations.

Guidelines for a Healthy Relationship

Is the situation the same in all industries? Is it in fact acceptable? It may be argued that where the sponsor is attempting to influence the sponsored person to buy a product for personal use, at personal cost, and at personal risk, sponsorship is an acceptable business practice. However, where the sponsor is attempting to influence the sponsored person to use a product on another person, at risk to that person and at the cost of a third party, the situation seems clearly different. Such a situation is more closely parallel to the bribes for letting contracts that are a generally condemned practice when accepted by Government or company officials. If we accept that sponsorship or gift giving does influence choice of product when confronted by a patient, the doctor must be found guilty of accepting a bribe (Wolfe, 1991). This should be labelled unethical behaviour. The American Medical Association has recently adopted guidelines which state that 'Any gifts accepted by physicians individually should primarily entail a benefit to patients and should not be of substantial value (Greenberg, 1990).

Accordingly, textbooks, modest meals, and other gifts are appropriate if they serve a genuine educational function.'

It is argued that these guidelines are sound (Greenberg, 1990) but the size of the gift merely defines the cost that one puts on one's integrity, not the level of integrity. There can be no justification in the individual accepting such gifts given that they are designed to, and do, influence behaviour (Thomson, 1990c; Thomson, 1990d).

In the case of organizations accepting support from the industry there are similar dangers (Greenberg, 1990). If the organization is dependent on such assistance, the industry may exert controls over the organization in a way which may influence patient care. To the extent that the medical profession has input into the decisions of organizations involved with patient care it should endeavour to prevent dependence on, or acceptance of, industry support. A similar argument may be raised against the acceptance of pharmaceutical funding for university teaching and research posts. Such posts are becoming more common as medical schools are being encouraged to move towards the new commercial ethic. While there is no evidence of direct or indirect harmful effects resulting from the acceptance of such sponsorship, equally there has been no research aimed at measuring possible ill effects.

What guidelines should be adopted by the profession if they are to keep our dealings with the industry on an ethically sound footing? I suggest the following:

(1) No individual practitioners should accept gifts of any kind from the pharmaceutical industry. This would include items ostensibly for patient benefit, such as trial or starter packs of medication, or subsidized literature.

(2) No organization which is involved in patient care should accept gifts of any kind from the pharmaceutical industry. Payment for services rendered, such as medical research, would be acceptable as long as the research fell within agreed guidelines (*vide infra*), and the level of payment was of the same order as would be made by any other non-pharmaceutical company for a similar level of time and expertise.

(3) Medical practitioners should not be involved in any education programme, either for the public or the profession, that links or could be seen to link, the pharmaceutical industry with the profession.

(4) Medical research which is sponsored by pharmaceutical companies should not be accepted unless it meets certain ethical guidelines:

(a) Where the research does not use a non-pharmacologic comparison there must be a case supported by scientific evidence to show that non-pharmacological treatment would not be in the best interests of the patient(s).

(b) There must be no restrictions on data analysis and publication independently of the company.

(c) The guidelines of the Declaration of Helsinki (Basch, 1990) and other local ethical guidelines designed to protect the patients' interests must be met.

These guidelines would not apply to research done by medical practitioners who were employees of a pharmaceutical company where meeting the requirements of the Declaration of Helsinki would be considered sufficient.

Such a set of ethical guidelines would probably be met by a storm of protest. Claims that these would penalize patients by reducing the care available to them, and doctors by reducing their access to education, would very likely be made. The former claim is not tenable since the cost of the industry's subsidies of patient care are met by an increase in the cost of pharmaceuticals. The latter would be true, however, if doctors insisted on the current over-priced and over-serviced education. A reduction in expectations relating to the non-educational aspects of courses is needed. In addition, the medical profession should recognize that its current behaviour creates educational bias and increases the cost of the pharmaceutical bill.

While there is the potential for these guidelines to reduce overall research, the lack of competition for resources by the 'industry biased' research could mean a better balance of research. An increase in interest in non-pharmacologic managements would also be a desirable outcome.

But what of the free journals that doctors receive in the mail? — journals influenced by, and full of advertising by the industry. These cannot be stopped. Nor in fact can any other mailed give-aways such as pens, calendars, pads etc. One may be able to reduce the effect of the mailed give-aways by making the above guidelines clear, and educating doctors about the reasons for the guidelines. It is possible that doctors might dispose of such give-aways rather than using or displaying them. Unfortunately, even in opening such mail and reading enough to decide to dispose of the items, the general practitioner will have already fulfilled the majority of the companies' aims! However, the journals fulfil a particular need for pharmaceutical and other information. Perhaps only if such information were sponsored by another source, or produced by the Government, would the need for the industry-sponsored journals cease to exist. In the current climate of Governmental spending constraints, and the unlikely event of finding non-industry sponsors, the journals are likely to remain. Producing these journals on a subscription basis with the full cost being met by practitioners is unlikely to be viable. If a journal was produced that was not dependent on pharmaceutical sponsorship it is likely that the industry would attempt to maintain an educational publication, free to doctors, that would carry its message. Such competition, in addition to the competition from other forms of medical education, would result in a paid subscription base too small to allow an independent review journal to be published at a price acceptable to the doctors while maintaining a sufficient profit margin for the publishing company. An added concern is that increases in the price that doctors pay for their education will eventually be met by higher patient charges. This will increase the annual cost of medical care, particularly for those who can least afford it. One 'advantage' of medical education being subsidized by the pharmaceutical industry has perhaps been that this cost is eventually met by the tax-payer, and hence met differentially by ability to pay!

The only strategies that may succeed therefore lie in trying to remove the apparent editorial necessity to produce articles that sell advertising by their content. Methodologies that enhance the status or importance of these journals, while ensuring their contents are based on the needs of practitioners, must be explored. Such strategies would ensure advertising continuance while reducing the bias in context. The problem of advertising's effect on prescribing, however, would not be solved. Likewise it is hard to see it becoming unethical to accept a visit from an industry salesperson.

Despite such difficulties, the guidelines proposed above would send a clear message to the profession. If continually reinforced, and if enforced by the professional bodies, they would produce a gradual change in perception. Such a perceptual change may also reduce the subtle influence of advertising on prescribing by making doctors look at their prescribing behaviour more critically.

It is unlikely that legislation would be able to be framed that would effectively end the undesirable influence of the pharmaceutical industry on the prescribing behaviour of the medical profession. The medical profession has developed self-regulation through ethical guidelines which, while imperfect, are usually more effective than complex legislative controls. However, the current guidelines have been shown to be inadequate to deal with the problems that may arise in the relationship between the pharmaceutical industry and the medical profession. New guidelines, such as those proposed above, must be developed and enforced. This is unlikely to occur without a will within the profession, and perhaps public pressure. Thus the major obstacle is the failure of the medical profession to recognize that they may be adversely influenced by the pharmaceutical industry. As intelligent individuals they believe that they can be uninfluenced by the industry's marketing. This has been clearly shown to be false. Nevertheless, more research should be carried out documenting the effects of pharmaceutical marketing. Education of the profession about the problem is critical if change is to occur, and producing a significant attitude and behaviour change requires strong evidence and continual reinforcement. Attitude change is not easy to produce, and tends to occur slowly. It may require more ongoing support than is needed to produce knowledge base change, and will be most successful if the impetus for change comes from within the profession. Funding for this educational endeavour will be a problem; certainly the industry would be reluctant to fund it, and the profession may be unlikely to put significant funds into a process likely to increase their costs! Some governmental assistance may be essential if an effective programme is to be mounted.

There is a close and essential relationship between the medical profession and the pharmaceutical industry. However, each has a different role to play and has a fundamentally different ethical responsibility. The ethical code of the medical profession fails to adequately deal with this relationship. Change in this code is urgently needed.

References

Avorn, J., Chen, M., and Hartley, R. (1982) Scientific versus commercial sources of influence on the prescribing behaviour of physicians *American Journal of Medicine* 73: 4-8.

Barham, P.M. and Benseman, J. (1981) *The general practitioner as a lifelong learner*, University of Auckland.

Barham, P.M. (1991) Personal communication.

Declaration of Helsinki (1990) in Basch, P.F. (ed) *International Health* pp. 410-13 New York, Oxford University Press.

Durham, J. (1990) Pharmaceutical promotion in New Zealand — a view from General Practice. In *Proceedings of the Public Health Association of New Zealand Workshop*: 43-9, Public Health Association.

Fogel, M.L. (1989) Survey of pharmaceutical promotion in a family medicine training programme *Canadian Family Physician* 35: 1063-5.

Greenberg, D.S. (1990) Washington perspective *Lancet* 22-29 December: 1568-9.

Hemminki, E. (1977) Content analysis of drug-detailing by pharmaceutical representatives *Medical Education* 11: 210-15.

Holmes, J.D. (1977) Prescribing information in drug advertisements *New Zealand Medical Journal* 86: 467-9.

Kawachi, I. (1990) Pharmaceutical advertising and promotion — options for action (Editorial): 1-20. In *Proceedings of the Public Health Association of New Zealand Workshop* Wellington.

Lloyd, J.S. and Abrahamson, S. (1979) Effectiveness of continuing medical education: a review of the evidence *Evaluation and the Health Professions* 2 (3): 251–80.

McLauchlan, W. (1991) *New Zealand Pharmacy* 11(3): 2–3.

McLauchlan, W. (1990) For the price of a pen *New Zealand Medical Journal* 103: 250.

Murdoch, J.C. (1990) For the price of a pen *New Zealand Medical Journal* 103: 190.

New Zealand Board of Health (1973) *Drug dependency and drug abuse in New Zealand, second report* (Report No. 18) Wellington.

New Zealand Council for Postgraduate Medical Education (1983) *New Zealand Medical Journal* 96: 578–9.

New Zealand Department of Health (Health Benefits Section) (1991) Personal communication.

New Zealand Drug Tariff (1990) Wellington, New Zealand.

New Zealand Medical Association (1990) *Code of Ethics* New Zealand Medical Association, Wellington.

Pharmaceutical Manufacturers Association New Zealand Inc. (1984) *Code of Practice* Pharmaceutical Manufacturers Association.

St George, I. (1989) The competitions (Editorial) *New Zealand Family Physician* 16: 186–7.

Thomson, A.N. (1990a) Unpublished data.

Thomson, A.N. (1990b) General Practice Research and the Pharmaceutical Industry *New Zealand Family Physician* 17: 195–7.

Thomson, A.N. (1990c) For the price of a pen *New Zealand Medical Journal* 103: 278–9.

Thomson, A.N. (1990d) For the price of a pen *New Zealand Medical Journal* 103: 135.

Wolfe, S.M. (1991) Promotional practices in the pharmaceutical industry *HAI (Health Action International) News* 59: 1–11.

Patents and Generics: A Campaign Diary

Joel Lexchin and Pauline Norris

If there is one issue that epitomizes the position of the pharmaceutical industry in its relationship with science, the professions, the consumer, and the Government, it is that of generics and patent protection. Research-based pharmaceutical companies spend a great deal developing and testing new drugs. Companies take out patents to protect this investment they have made. While a new drug is covered by a patent, no other companies can manufacture it. This means that research-based companies have an interest in keeping the patent length as long as possible. When the patent expires, other companies can manufacture their own brands of the drug using the generic formula (generics). Companies which manufacture generics face both prescriber brand loyalty and originator company attempts to discredit their products. These hinder their attempts to market their products, which could otherwise lead to competition and savings for consumers. This chapter describes the scientific, professional, and commercial issues and shows how the industry works to protect its position.

Patents — Rights and Wrongs

The question of the appropriate patent life for a drug is sure to draw a response from the Researched Medicines Industry (RMI), the association representing the multinational drug companies. The response is always that the patent life is too short. The reason for this predictable response is simple. While a drug is under patent protection, no one else can sell it; therefore, the longer the patent, the longer the company has a monopoly on the drug, and the greater the profits. In New Zealand, patents are valid for 16 years after the date of application, but they can be extended for up to another 10 years.

As well as keeping prices, and therefore profits, high, patents encourage competing pharmaceutical companies to market 'new' drugs which are only slight alterations on the patented drug. The changes only have to be sufficient to warrant a new patent; no therapeutic advantage is needed.

These 'me-too' drugs create further confusion for prescribers. Kawachi and Lexchin point out in their chapter that most drugs that come onto the market demonstrate little or no therapeutic advance over existing ones.

The RMI gives altruistic reasons for wanting patents strengthened. It maintains that 'patents are a well recognized vehicle for the transfer of technology. By offering protection of new products and processes they encourage capital investment in research and development' (Scott, n.d.). However, the industry spends only $1 million annually on research in New Zealand (Pharmaceutical Manufacturers Association (PMA), 1987) compared to $19 million in Denmark (Burstall et al., 1981), a country of comparable size. Even in some industrialized countries with stronger patent protection the multinationals still do not invest heavily in research (Burstall et al., 1981). The industry's stock answer is that it can't do research here because of weak patent protection (PMA, 1984).

The necessity for patents can also be challenged at a more fundamental level. Dr Alan Klass, who conducted a wide-ranging inquiry into the pharmaceutical industry for the Canadian province of Manitoba and then popularized his findings in the book *There's Gold in Them Thar Pills*, argues that from the point of view of scientists, patents are a foreign concept. They developed not because of anything within the nature of science, but purely for commercial reasons. Patents can impede the free flow of ideas, the life blood of science. Once a discovery is patented, other scientists can use it only with the permission of the patent holder. Yet no scientific discovery stands by itself; it is built upon the work of other researchers. If the person who discovered the method of growing polio virus had patented his idea instead of freely publishing it, then the polio vaccine might never have been developed. As Klass concludes: 'It was only when corporate interest became dominant that the right of a party claiming a patent for the discovery became prominent. Patents serve the industry much more than the individual discoverer and certainly much more than society' (Klass, 1975).

For the past half dozen years the RMI has been waging a vigorous campaign to get the Government to lengthen the period of patent protection to 20 years, with an option for a further five years. One past executive director of the association warned that a failure to extend the patent life might result in many new drugs being withheld from New Zealand because manufacturers may find it uneconomical to market their drugs here (Gasson, 1984). Periodically, the RMI repeats the same threat (PMA, 1987), but when one of its representatives was directly questioned on this point he could not name a single drug or company that had actually carried through with the threat (Radio New Zealand, 11 May 1987).

As part of its battle plan, the RMI has run major national media campaigns under themes such as 'medicines are the biggest bargain in New Zealand's health care system' (Anon., 1985). There have been numerous briefs to Government (Anon., 1985; Stockdill, 1987; PMA 1988), all carrying the same message.

One of the RMI's main arguments for longer patents has been the assertion that by the time a drug has gone through all the required testing, been approved for marketing, and been accepted for inclusion in the Drug Tariff, there may be as little as five years left on its patent (PMA, 1984; Anon., 1985). The RMI companies contend that this remaining time is insufficient to recover their heavy investment in researching and bringing their products to market. Malcolm Eppingstall, a past president of the RMI, called attention to the high costs involved: in the order of $100 to $200 million (Gasson, 1984).

The $200 million figure may, actually, be an exaggeration. A recent preliminary report from the United States (US) Office of Technology Assessment has questioned this estimate and describes the amount as 'an arbitrary number with no intrinsic meaning' (Anon., 1991). However, whatever the true cost of bringing out a new drug, the line of argument used by Mr Eppingstall leaves out one crucial fact. When the RMI gives its figures, they are not referring to New Zealand costs, but to costs worldwide. Since New Zealand represents about 0.2 per cent of the world pharmaceutical market, the New Zealand share of development costs that need to be recouped is in the order of $200,000 to $400,000. Even so, the multinational companies are still not willing to concede that they are recovering their expenses (Stockdill, 1987). If this were the case, one would expect it to show up in low profit figures; yet the pharmaceutical industry has consistently been among the most profitable of all manufacturing industries in New Zealand (Lexchin, 1989).

In all probability the RMI has a strong commercial motivation; it wants to extend patent protection to boost profits even further. Patent protection is a major factor in keeping drug prices high, and if prices are high, so are profits. When a New Zealand parliamentary subcommittee on the cost of drugs reported in 1968, it recommended that 'urgent consideration be given at this juncture to *reducing* the period for which patent protection is given from 16 years to 10 years in the case of drugs [emphasis added].' The subcommittee found that patents were probably keeping up the cost of drugs. There was little voluntary reduction in the price of most drugs that were subject to patent protection even where sales were comparatively high (Public Expenditure Committee, 1969).

It is also worth considering what causes such a long delay between the time a patent application is filed and the time the drug appears on the New Zealand market. For drugs originating in the US, and many drugs here come from American companies, nearly all the delay occurs in the US — almost 10 years. This is the time that it takes for development and approval of the drug in that country. Only about one and a half years of the delay is due to the time taken for approval of the drug in New Zealand (Parker, 1985). The development and testing in the US is vital for the drug's approval in New Zealand because these are the tests that the Medicines Assessment Advisory Committee uses to make its recommendation about whether the drug should be licensed in New Zealand. Therefore, in essence, the RMI's position is that patent life in New Zealand should be extended because things take so long in the US.

Generics

If the RMI companies are already recovering the share of development costs incurred in New Zealand and if they are not going to expand the amount of research they do in this country, then one has to wonder why there is the sustained campaign over patents. The reason is undoubtedly the challenge to industry dominance posed by generics. RMI companies are, as the name suggests, research-based companies. They develop new drugs and market them under their own brand name. After their patent expires other companies can manufacture their own brands of a drug and compete with the originator company. These are usually referred to as 'generics' and companies which specialize in manufacturing them are known as 'generic manufacturers'. For a variety of reasons doctors usually continue to prescribe by original brand name after the patent expires. The brand name may have become so synonymous with the product that this happens quite unconsciously — think of 'aspirin', which was originally a Bayer brand name (McTavish, 1987) or 'Hoover', a brand name for vacuum cleaners which people now use as a general name. Aside from established position, however, a major reason for this 'brand loyalty' among doctors is anxiety about the quality of generics, an anxiety that is subtly and constantly fuelled by the research-based companies.

Sales of generics have been expanding in New Zealand, from 6 per cent of the market in 1982 (Anon., 1983) to 20 per cent in 1987 (New Zealand Medical Association (NZMA), 1987). But the multinational members of the RMI are not just concerned about the loss of 20 per cent of the New Zealand market. They are also worried because the availability of a generic equivalent could force them to lower their prices. If a lower-priced generic enters the market that has hitherto been dominated by a drug produced by one of the RMI companies and listed on the Drug Tariff, then the Department of Health may impose what is known as a part-charge on the brand name product. The part-charge is the difference in price between the generic and brand name versions of the drug in question. The Department of Health only pays the pharmacist the price of the generic drug, regardless of which version is dispensed. If it is the brand name product, then the patient has to pay the difference (the part-charge). Once patients and doctors know that certain medications carry a part-charge, the doctors may switch, or the patients may ask them to switch, to the generic product. Therefore, unless the RMI companies want to see their share of the market for those drugs decline, they may be forced to lower their prices towards those of their generic competitors.

One fact often overlooked in this debate is that multinational research-based companies have themselves been moving into the generic manufacturing field over the last 10 years or so. This means that companies are, in effect, increasingly arguing against the use of their *own* products — which are, of course, cheaper. It also means that small generic producers may be driven out of the market, which will then become dominated by the multinationals.

It is important to distinguish 'branded generics', those manufactured by the large corporations and some others manufactured under brand names once the patents have lapsed, and 'commodity generics', those produced by smaller producers who do not use brand names and supply either the bulk demand of hospitals and other institutions, or operate in the mail order market in the US. In 1979 branded versions constituted 93 per cent of generics sales in the US. All 10 leading suppliers of branded generics were large transnational corporations — including Searle, Eli Lilly, Roche, Ciba, Upjohn, Squibb, and Smith Kline and French. Forty-two per cent of the *commodity* generics market was occupied by four transnationals (United Nations Conference on Trade and Development, 1981). This may allow multinationals to drive smaller, locally-owned generic companies out of business by tactically lowering prices to unprofitable levels. Once the smaller firms have disappeared, competition is reduced and the multinationals can increase prices again.

In New Zealand, some multinationals have set up generic subsidiaries. For example, in August 1984 Glaxo opened Evans Medical (NZ) Ltd to produce branded generics. However, multinational domination of the generics market is not inevitable. In Canada, the two largest generic companies are locally owned.

The Campaign Against Generic Prescribing

A related issue to that of generics is the question of how doctors should describe the drugs they prescribe — should they describe them by brand or by chemical (generic) name? Each drug has a unique generic name, but may be marketed by a variety of different companies under different brand names. The RMI would prefer it if New Zealand doctors did not use generic names when they prescribe. It commissioned a booklet entitled *The Case for Prescribing by Brand* that contended that 'by teaching prescribing according to brand, medical academics will encourage . . . good medical practice' (Scott, n.d.). In fact, many clinical pharmacologists — the doctors who specialize in how drugs work and should be used — recommend generic prescribing for some very sound reasons. First, the generic names will tell doctors that two drugs are related and probably will have similar effects. For example, from hearing the trade names you could hardly tell that Inderal, Visken, and Aptin were all related. However, the generic names make it obvious: propranolol, pindolol and alprenolol. These all come from a family of drugs for the treatment of angina. The second reason is that trade names can often hide the contents of a drug. If doctors do not know what is in the product they are prescribing, they may be making a serious error in prescribing. While thalidomide was still on the world market, a child with phocomelia (malformation of the limbs) was born in Brazil. Since thalidomide typically caused this type of malformation, the Brazilian *Ocruziero* magazine investigated, but was told that thalidomide was not on sale in Brazil. It was later determined that thalidomide was being sold under five different trade names in that country. None of these names provided any clue that the drugs contained thalidomide.

To further its campaign to stop generic prescribing, the RMI has warned that the small size and limited resources of generic companies means that they cannot provide all the information about drugs that the multinationals can (Scott, n.d.). However, a study in Canada has shown that there is no difference between generic and brand name manufacturers in the quality of information they provide to doctors, and that generic manufacturers respond faster than the brand name ones (Thomas and Lexchin, 1990).

Sometimes RMI members try a direct inducement to get doctors to prescribe by brand name. An advertisement in the *New Zealand Medical Journal* for Floxapen (Beecham's brand of the antibiotic flucloxacillin) announced that for every prescription written using the brand name, Beecham would make a donation to the New Zealand Neurological Foundation (Wright, 1983). Another tactic is to play on fear. Dramatic headlines such as 'Australians Object to Generic Prescribing' or 'British

Warning Against Generic Prescribing' have appeared in advertisements in the *New Zealand Medical Journal*. Another approach is for individual companies such as Allen and Hanburys, a division of Glaxo, to write letters to doctors asking them not to prescribe generically.

In trying to influence doctors against generic prescribing, the RMI has consistently questioned the quality of generic products. When a practitioner wrote a letter to the *New Zealand Medical Journal* querying the quality of generic equivalents to Lasix (Hoechst's brand of the water pill frusemide) (Bailey, 1979), the executive director of the PMA was quick to warn that 'the PMA is of the opinion that the Department of Health and the medical profession have no guarantee of equivalence when a generic product is marketed' (Martin, 1979). A subsequent study comparing Lasix and four generic versions found no significant differences between any of them (Eggers et al., 1983). Although the RMI now concedes that generics have a legitimate place in New Zealand, it is still keeping up its attack on their quality (Anon., 1990a). Yet, despite its best efforts, the RMI has never provided any convincing evidence that the use of generics has harmed the health of any New Zealander.

The threat of lower profits is why the executives from the New Zealand subsidiaries of multinational companies are so quick to denounce generic drugs and their manufacturers. Bob Williamson of Wellcome New Zealand Limited decried the 'Arab bazaar tactics . . . being used by some generic pharmaceutical companies' (Anon., 1982). R.W. Martin, a former executive director of the RMI, said that his organization 'considers that the Department of Health is playing with fire in seeking cheaper generic products . . . In both the short and the long-term, it is the patient who is at risk from such a policy' (Martin, 1979).

Generic Substitution

The issues of generics and generic prescribing came to a head in the question of substitution. At present New Zealand pharmacists are allowed to substitute drugs from a different manufacturer if an agreement is reached with the doctor and put in writing by them. In late 1988 David Caygill, the then Minister of Health, announced that a change would be made to the Medicines Regulations, to allow substitution, unless prohibited by the doctor or the patient. This would not, of course, affect drugs that are still under patent and for which, therefore, there were no alternative brands available. This is called generic substitution because it permits the interchange of drugs with the same chemical description but with different brands or manufacturers.

Experience in Canada has clearly shown that a policy of increasing the availability of generics and allowing them to be substituted for brand name drugs can lead to substantial savings. Drugs produced by two companies are about 20 per cent cheaper than those produced by a single manufacturer; when the number of companies rises to five, then the differential is almost 50 per cent (Lexchin, 1991). In 1983, Canada was conservatively saving $211 million out of a total drug bill of $1,600 million (Commission of Inquiry on the Pharmaceutical Industry, 1985). Recently Canada strengthened its patent law making it almost impossible to market generic drugs for the first 10 years after the introduction of a new drug. This change came about because of intense pressure from the US Government and the multinational pharmaceutical companies.

The US and Canada have gradually allowed generic substitution since the early 1960s. There was a burst of legislative activity in the late 1970s and the number of States in the US permitting substitution jumped from 24 to 45 in the two years from 1977 to 1979 (Rosenberg, 1979). The legislation varied from State to State and led to different degrees of success in reducing drug costs. There is evidence that in States where doctors simply had to sign on a different line to indicate whether substitution was to be permitted, the percentage of doctors opting against substitution was high (62 per cent and 75 per cent in two studies). The States which required physicians to write a comment such as 'Do not substitute' led to fewer vetoes (3.6 to 6.4 per cent in four studies). In the Canadian province of Ontario in 1989, doctors wrote 'no substitution' on only 1.8 per cent of prescriptions (Gorecki, 1990). In New Zealand's proposed regulations doctors would have had to write 'No substitution' on the prescription form.

The potential for competition among the best-selling drugs in New Zealand is enormous: almost 70 per cent of the 30 top-selling drugs are only available from one company. The previous Labour Government announced that it intended to increase competition through parallel importing and encouraging generic substitution. The latter would require a change in the Medicines Regulations allowing pharmacists to dispense generically unless the doctor or the patient specifically asked otherwise. It has been estimated that together parallel importing and generic substitution could have reduced pharmaceutical expenditures by about $50 million annually (Sinclair and Beaglehole, 1989).

In New Zealand, as elsewhere, the introduction of substitution has prompted a great deal of outcry from the research-based manufacturers. Allowing pharmacists to substitute could seriously undermine the market impact of brand loyalty. In other words, the companies could still successfully maintain brand loyalty among doctors but fail to prevent cheaper generics from gaining an increased market share. When the initiative was first announced in September 1988 the organization representing the industry in New Zealand promptly changed their name from the Pharmaceutical Manufacturers Association to the Researched Medicines Industry (RMI) and embarked upon a public campaign emphasizing the research basis of their products. The move also permitted them to exclude generic manufacturers and importers from their organization.

For obvious reasons, the RMI was opposed to these proposals, 'on scientific, legal and commercial grounds' (PMA, 1988). According to Bill McLauchlan, then chief executive officer of the RMI, the Government 'appeared hell-bent on portraying the pharmaceutical industry as an enemy of the people' (Anon., 1990c). The RMI geared up for a major campaign.

Even before this proposal was announced, the industry was active in trying to prevent illegal substitution. In 1987 a pharmacist was charged with this offence in Wellington. Although the case failed on a technicality, it created something of a storm and indicated that substitution was an issue of continuing significance. The pharmacist's lawyer claimed that 'pressure to prosecute had been put on the Health Department by drug manufacturers' (Moffett, 1987).

The Campaign Against Substitution

After the 1988 proposal to allow substitution without the explicit consent of the doctor, the research-based companies stepped up their campaign. The *Christchurch Press* reported the PMA warning that this could encourage big drug companies to shut down their New Zealand operations. The president of the PMA said that the availability of generics threatened the viability of company research and could put an end to their funding of university and hospital research programmes. He also had 'very real concerns' about the quality of some generics. The *Press* also published the Health Department's response. The manager of the Primary Health Care Unit of the Department of Health, Dr Bob Boyd, argued that New Zealand may have been paying more than its share of research costs in the past, that the safety issue was 'scaremongering' and that drug companies had been threatening to leave the country for years but had not done so. He also said that generic drugs were quality-tested by the Health Department before they were allowed into New Zealand.

Possibly the most expensive item in the industry's campaign on this issue was a 30-minute RMI television programme screened on a Saturday morning in March 1989 called 'A Rose by Any Other Name'. This included interviews with doctors and industry representatives. There was quite an extensive interview with a cancer specialist from the Auckland Area Health Board. He said that his patients had experienced significant side-effects from using a generic drug. The programme also concentrated on the reports of a Johnsonville doctor who claimed that one of his patients had experienced uncontrolled high blood pressure while on a generic medicine. It was claimed that the patient's blood pressure had dropped when transferred to the innovative brand. Both these cases were disputed by the Health Department, although not publicly. In the first case it appeared that an unapproved drug was being used, and in the second the batch number the pharmacist had given corresponded to a drug other than the one prescribed as the blood pressure medication (Department of Health, 1989). This programme was aimed at the medical profession but its screening was mentioned in the *Dominion* of two days before (9 March 1989) and comments in the media, especially on radio talkbacks, suggest that many other people, mostly the elderly, also watched.

In April 1989 the RMI called for the Minister of Health to withdraw eight generic drugs distributed in New Zealand which were said to have been withdrawn in Australia the previous September. They claimed that 25,000 New Zealanders were on these drugs, which included antibiotics, treatment for heart disease, and anti-diabetic products. Links were explicitly drawn with the issue of generic substitution — the RMI spokesperson claimed that 'if the Government continued with plans to introduce substitution of a doctor's prescription with a medicine of a pharmacist's choice, up to 80,000 New Zealanders could be put on to one of the products withdrawn in Australia' (*Evening Post*, 21 April 1989). RMI chairman, Murray Main, raised questions about the safety of the 'hundreds of generic products' used in New Zealand, and accused the Health Department of taking a cavalier attitude to the dangers of inferior drugs in a rush to cut down the drug bill — which he said only amounted to 25 cents a person a day (at manufacturers' level). The Health Department issued a press release pointing out that the medicines named by the RMI were made in a New Zealand factory to a New Zealand formula, whereas those recalled in Australia were made in that country to an Australian formula. David Pickering of the Medicines and Benefits section of the Department said that 'the issues which led to their withdrawal in Australia are simply not relevant to the New Zealand products'. He did note, however, that the Department had, that day, ordered the recall of a drug manufactured by an RMI member company, because of the presence of an illegal colouring agent!

The RMI again took the issue to the public — this time in a series of television, newspaper, and magazine advertisements. These showed an empty tuberculosis sanatorium and claimed that 'thanks to research funded by the medicines industry' diseases such as tuberculosis, diphtheria, scarlet fever, and polio are no longer a problem and that if the industry continues its good works, 'we may be able to close down the cancer wards too'. This bore some resemblance to an earlier advertisement run in the *New Zealand Medical Journal* (*NZMJ*) some years ago which showed gravestones in an attempt to persuade doctors to 'Support Research — Prescribe by Brand Name' (*NZMJ*, 26 June 1985). These arguments ignore the fact that the dramatic decline in mortality rates from infectious diseases is due less to drug therapy than to late nineteenth century improvements in water supply, sanitation, nutrition, and housing (see for example, McKeown, 1976; Taylor, 1979).

In September 1989 *Eyewitness News* reported that moves to allow substitution were continuing despite recent concern in the US over the safety of generics. It claimed that faked safety tests and bribes made to Federal officials by generic companies had recently come to light. It was reported that out of 12 generics tested by the US Food and Drug Administration (FDA), 10 had failed to meet standards. In New Zealand, Pfizer, who the programme said had reported 'pretty average profits', were fighting hard against substitution. Doctors, who *Eyewitness News* said, 'have no real axe to grind', were also fighting the initiative for safety reasons. Two doctors, who also appeared on the RMI video, recounted stories of adverse reactions to generics. The then Minister of Health, Helen Clark, reported that the Government was determined to press ahead with the initiative, that individuals vary greatly in their reactions to drugs, and that Health Department procedures would ensure that generics would be as good as brand name products.

Two months later, a *Foreign Correspondent* television programme gave more information about the US 'scandal' mentioned in *Eyewitness News*. It had been shown that an official in the generic drug division of the FDA had been accepting bribes, and generic companies had been submitting false information to the FDA. This was exposed by a private investigator hired by a rival generic company. As a result of his findings 29 drugs were recalled and 150 put on hold. This represented 2 per cent of the generics on the US market. Explicit links were drawn between this scandal and generic substitution in New Zealand — the presenter said that some New Zealand doctors see this scandal as 'a strong warning about what could happen here'. Neither the introduction nor the programme gave any evidence that corruption was more likely to occur in the generic sections of regulatory agencies, or that generic companies are more likely to submit misleading information.

Later evidence suggested that the scandal did not in fact endanger public health. The FDA found that applications from firms who had bribed their officials had not been approved any faster than others, and that drugs had not been approved which did not merit it. The FDA also launched a major investigation of generic drugs, which had reassuring results. Only one per cent of more than 2500 samples of popular generic drugs tested for purity, potency, dissolution rate, product identity and so on did not comply fully with drug standards. This is consistent with normal rates for both generic and brand name drugs. Even when drugs with a narrow therapeutic range were tested, such a low error rate was found that the FDA considered that there was no health risk (Anon., 1990a).

Impact of the Campaign

The Department of Health appears to have been slow in responding to some of the publicity from the industry and other groups. Often claims made in newspapers, on television, and on radio were not countered by the Department, and sometimes the responses that reached the public were not very clear. The RMI's campaign was greatly assisted by their sponsorship of New Zealand journalists under the PMA Journalist Award scheme.

The RMI campaign against generic substitution may be having an effect on New Zealand doctors. Although two-thirds of those who responded to a recent survey prescribed generic medicines when appropriate, a majority were against allowing pharmacists to substitute generic drugs for brand name ones. A substantial number were also concerned about the quality of generic medicines. The authors concluded that:

Seventy-eight cases of problems encountered with generic medication were provided. Most of the cases can only be regarded as anecdotal. There is a lack of documentary evidence to support the views that prescribing of generic drugs constitutes a threat to patient safety. With regard to the problems experienced with generic medication, it is recognized that a similar questionnaire seeking a record of problems experienced with brand-named medications may provide a similar response (Tilyard et al., 1990).

The safety concerns of this group of doctors echoed the view of a working group of the NZMA. In its report the working group called for clear Government guidelines on bioequivalence. If generic drugs are to assume a wider role in New Zealand it is important to gain the confidence of the medical profession, and guidelines such as those proposed by the NZMA seem reasonable.

The Department of Health has always maintained that the quality of generics is as good as that of the brand name drugs. In routine testing of drugs the number of recalls among products from RMI companies is about the same as from generic manufacturers (Division of Clinical Services, 1979). According to one Government official, on a number of occasions the generic product slightly out-performed the brand name product in bioequivalence testing (Anon., 1983). Once again, the RMI could not provide any solid evidence to substantiate its claim about the inferiority of generic drugs. Even the so-called scandal a few years ago about improper testing of generic products by a South African laboratory never identified any specific problems with generic drugs (Stockdill, 1987).

Individual companies within the RMI have gone to court to delay the appearance of generic drugs. Douglas Pharmaceuticals, one of New Zealand's largest generic manufacturers, had imported sample quantities of generic cimetidine and supplied them to the Health Department to support a marketing application in anticipation of the expiry of the patent. Smith, Kline and French, which owns the patent, sued, claiming that this action was an infringement on their patent, and won in the Court of Appeal (Anon., 1991). As a result, it will probably take even longer for generic products to appear in New Zealand, and without competition prices will stay high for much longer.

Conclusion

Studies have consistently shown that drug prices in New Zealand are high by international standards. The Coopers and Lybrand (1986) study on removing medicines from price control reported on two international comparisons of drug prices. In one sample of 22 medicines, using 1984 prices in six markets, New Zealand had 55 per cent of its medicines higher than the sample average, and in 23 per cent of cases had the highest price in the sample. In the second study of five markets for 40 medicines, using 1986 prices, 48 per cent of the medicines were more expensive in New Zealand than the weighted average (Coopers and Lybrand, 1986). Another study by the Australian Industries Assistance Commission of nine out of the 10 most commonly prescribed medicines in Australia and New Zealand suggested that New Zealand prices were higher by some 37 per cent (Scott et al., 1986).

There are a wide range of options for controlling public expenditure on drugs (Sinclair and Beaglehole, 1989). But some of them, such as increasing patients' co-payment or not including new, more expensive products on the Drug Tariff, are socially regressive and can penalize the poor and the sick. On the other hand, increasing competition by encouraging the use of generics has almost no down-side effects, even for the multinational subsidiaries. The Canadian Commission of Inquiry into the pharmaceutical industry showed that the industry continued to thrive even with strong generic competition (Commission of Inquiry on the Pharmaceutical Industry, 1985). Generic drugs are not the only solution to limiting the drug bill, but they will help people to afford the medicines they need without restricting access to innovative drugs.

References

Anonymous (1982) Medicine makers warn against industry infighting *New Zealand Pharmacy* 2: 36.

Anonymous (1983) Govt spells out generics policy *New Zealand Pharmacy* 3: 17.

Anonymous (1985) Evans generics could prompt price war *New Zealand Pharmacy* 5: 15.

Anonymous (1985) Medicine Makers Campaign Launched *Marketing Magazine* 4: 68.

Anonymous (1990a) Generic Drugs: Still Safe? *Consumer Reports* 310-13.

Anonymous (1990b) New Zealand drug expenditure to be cut *Scrip* 1561: 19.

Anonymous (1990c) Storm brews over NZ import powers *Scrip* 1561: 21.

Anonymous (1991) SK&F wins NZ cimetidine patent case *Scrip* 1608: 16.

Anonymous (1991) OTA queries industry R and D cost data *Scrip* 1614-5: 20.

Bailey, R. (1979) Do cheaper generics equivalents contain what we think? *New Zealand Medical Journal* 89: 98.

Burstall, M., Dunning, J., and Lake, A. (1981) *Multination enterprises, governments and technology — the pharmaceutical industry* Paris, Organisation for Economic Co-operation and Development.

Commission of Inquiry on the Pharmaceutical Industry (1985) *Report* Ottawa, Supply and Services Canada.

Coopers and Lybrand associates (1986) *Removal of medicines from price control* Wellington, Department of Health.

Department of Health (10 March 1989; 15 March 1989) Memoranda.

Division of Clinical Services (8 June 1979) *Medical programme* Clinical Services Letter (No 187) 2.

Eggers, N., Saint Joly, C., Jellet, L., and Shirkey, R. (1983) Discrimination between five oral frusemide preparations using renal chloride excretion and frusemide recovery *New Zealand Medical Journal* 96: 963-6.

Fedder, D.O. (1971) Letter to the editor *Journal of the American Pharmaceutical Association* NS11, 12: 641.

Gasson, W. (1984) New Zealand's drug dilemma: the best or the cheapest? *Better Business* 48: 8.

Gorecki, P. (1990) *Getting it right: an evaluation of alternative systems of the organization of the Ontario prescription drugs distribution system* Toronto, Study for the Pharmaceutical Inquiry of Ontario.

Gray, B. (8 April 1989) Medicine Makers Campaign: Letter to the Editor *Listener*.
Griffith, R. (1987) Bioavailability in New Zealand *New Zealand Pharmacy* 7: 10.
Klass, A. (1975) *There's gold in them thar pills* Middlesex, Penguin.
Lexchin, J. (16 April 1991) More competition leads to lower drug prices (letter) *Medical Post* 15.
Lexchin, J. (1984) *The Real Pushers: a critical analysis of the Canadian drug industry* Vancouver, New Star Books.
Lexchin, J. (1989) Pharmaceutical Promotion in New Zealand *Community Health Studies* 12: 264-72.
Martin, R. (1979) Diuretic equivalence of frusemide preparations *New Zealand Medical Journal* 90: 209.
McKeown, T. (1976) *The Role of Medicine: Dream, Mirage or Nemesis* London, Nuffield Provincial Hospitals Trust.
McTavish, J.R. (1987) Aspirin in Germany: The Pharmaceutical Industry and the Pharmaceutical Profession *Pharmacy in History* 29: 3.
Mitchell, E. (12 March 1989) Generic drugs TV war flares *Sunday Star*.
Moffett, W. (1987) Substitution: Name dropping can offend *New Zealand Pharmacy* 7: 6-15.
New Zealand Medical Association (NZMA) (1987) *Pharmaceutical Funding in New Zealand: the costs, problems and a possible alternative* Wellington.
Parker, J. (1985) The effective patent life of American originated drugs in New Zealand *New Zealand Economic Papers* 19: 61-8.
Parker, L. and Field, M. (12 March 1989) Medicine makers put case on TV *Dominion Sunday Times*.
Pharmaceutical Manufacturers Association (PMA) (1984) Medicine Manufacturers seek end to part charges *New Zealand Pharmacy* 4: 26-7.
Pharmaceutical Manufacturers Association (PMA) (1987) *Medicines and Health: the pharmaceutical industry — improving the quality of life* Wellington, Pharmaceutical Manufacturers Association.
Pharmaceutical Manufacturers Association (PMA) (1988) Health Industry News *New Zealand Medical Journal* 101.
Public Expenditure Committee (1969) *Report* Wellington, Government Printer.
Rosenberg, C.L. (1979) How many patients still get the drug you prescribe? *Medical Economics* 56: 35-54.
Sands, S. (13 March 1989) Cheap drugs 'may threaten patient safety' *Christchurch Press*.
Scott, C., Fougere, G., and Marwick, J. (1986) *Choices for Health Care* Wellington, Health Benefits Review.
Scott, G. (n.d.) *The Case for prescribing by brand* Wellington, Pharmaceutical Manufacturers Association.
Sinclair, B. and Beaglehole, R. (1989) Strategies for reducing the drug bill *New Zealand Medical Journal* 102: 165-7.
Stockdill, R. (10 May 1987) Copy-cat drugs *Auckland Sunday Star* A12.
Taylor, R. (1979) *Medicine out of control: the anatomy of a malignant technology* Sun Books, Melbourne.
Thomas, M. and Lexchin, J. (1990) Pharmaceutical Manufacturers responsiveness to physicians' requests for information: a comparison between brand and generic companies *Social Science and Medicine* 31: 153-7.

Tilyard, M., Dovey, S., and Rosenstreich, D. (1990) General practitioners' views on generic medication and substitution *New Zealand Medical Journal* 103: 318–20.

United Nations Conference on Trade and Development (1981) *Trademarks and Generic Names of Pharmaceuticals and Consumer Protection* Report by the United Nations Conference on Trade and Development secretariat TD/B/C.6/AC.5/4.

Wright, L. (1983) Pharmaceutical advertising *New Zealand Medical Journal* 96: 407.

Doctors and the Drug Industry: Therapeutic Information or Pharmaceutical Promotion?

Ichiro Kawachi and Joel Lexchin

Introduction — the Prescribing Process

Chances are that the next time you visit your doctor's surgery you will leave with a prescription — over six out of 10 visits end this way (Simpson et al., 1984). We all hope that our doctors know what they are doing when they give out a prescription, but there are 8000 different medicines containing 2000 different active substances on the New Zealand market (Division of Clinical Services, 1980). With the time and energy demands of running a surgery it seems improbable that any doctor would have more than a passing knowledge of a small fraction of the medications available. Yet each year the list of new drugs grows faster than the rate of retirement of outdated medicines. The development of new medicines proceeds at a much faster pace than the growth of knowledge about existing ones. No sooner does the prescriber get used to a particular medicine than a new therapeutic alternative becomes available (Maling, 1989).

This situation is not unique to New Zealand. In the United States, 348 new drugs were introduced by the 25 largest American drug manufacturers between 1981 and 1988. Of these, the Food and Drug Administration deemed 3 per cent (twelve drugs) as having 'an important potential contribution to existing therapies'; 13 per cent as having 'a modest potential contribution'; and 84 per cent as having 'little or no potential contribution' (Randall, 1991a). The basic problem of pharmaceutical formularies throughout the world is that there are too many indistinguishable drugs ('me-too' drugs) spread over many therapeutic categories (Wolfe, 1991). Since there are often no real differences between new drugs and existing drugs, differences must be created based on advertising and promotion.

The forces influencing the prescribing process are complex. Over a lifetime of practice, the prescriber develops preferences for certain medicines based on an interaction between core knowledge of clinical pharmacology, and personal experience of patients' responses (what could be termed 'therapeutic empiricism'). Factors which modify a practitioner's prescribing habits include patients' expectations, time and money constraints (for example, the tendency to write prescriptions to terminate a consultation), as well as the constant barrage of promotional drug information from pharmaceutical companies (Maling, 1989). Increasingly, prescribers are being exhorted to modify their medicines preference according to 'unbiased' assessments of the costs and effectiveness of alternative therapies. Tariff restrictions and the publication of the Health Department's Clinical Notes Series represent attempts by the New Zealand Government to modify the prescribing process (more on these later).

Unfortunately, the important question of how much doctors actually know about what they prescribe lacks a simple answer; there have been no New Zealand studies which have measured how much doctors know about the drugs they prescribe. Nonetheless, it is still worthwhile to look at where they get their information about medicines. The better the sources that doctors use to learn about drugs, the more likely they are to prescribe them well.

Prescribing information can be categorized into two types: commercial, that is, coming from the drug industry, and professional, that is, based on non-commercial sources of information. The question of which type of source prescribers predominantly rely on is not merely academic. Studies carried out in Belgium, Holland, the United States, and the United Kingdom have consistently found that doctors who are better prescribers make greater use of professional sources of information (Becker et al., 1972; Linn and Davis, 1972; Mapes, 1977; Haayer, 1982; Ferry et al., 1985; Blondeel et al., 1987). Conversely, the more reliant doctors are on commercial sources of prescribing information, the less rational they are as prescribers (Avorn et al., 1982; Peay and Peay, 1988). By 'rational prescribing', it is meant that the drug treatment is appropriate, effective, and safe for the symptoms/disease presented, that the prescribed dosage and duration is correct, and that the most cost-effective regimen has been chosen for the presenting symptom/disease (Haayer, 1982).

Professional Sources of Prescribing Information

There are currently only two sources of non-commercial prescribing information available to New Zealand practitioners on a regular basis. Both are funded by the Health Department. Since 1959, the Department has employed Visiting Professional Staff to make regular calls on medical practitioners in the country. At present, the visiting staff consist of two Visiting Practitioners and one Visiting Pharmacist. One Visiting Practitioner is based in Auckland. Over a two-year period all general practitioners and a representative number of specialists in private practice in the North Island are visited. The other Visiting Practitioner is based in Christchurch and covers the whole of the South Island, making visits on an annual basis to general practitioners and specialists. Both Visiting Practitioners provide information on Departmental policies related to medical practice, discuss issues of importance in medical practice and obtain information from doctors on areas of urgent medical need. These visits may also cover prescribing patterns and costs (Scotney, 1990).

The Visiting Pharmacist is based in Auckland and covers the northern half of the North Island. His discussions with doctors focus primarily on prescribing behaviour and drug costs. Until recently, no formal evaluation of these services had been carried out (Scotney, 1990). On the other hand, the early attitude of the drug industry to supplying doctors with a non-commercial source of information can be gauged from the report of the past director of the Clinical Services Division of the Department of Health. According to Dr A.W.S. Thompson, the industry was concerned that these Visiting Practitioners would 'combat' the promotional efforts of the companies, especially those of their detailers ('drug reps'), the men and women who are paid by the pharmaceutical companies to go from surgery to surgery promoting their products to doctors (Thompson, 1970).

With hindsight, it could be said that the pharmaceutical industry had little cause for concern. A national survey of general practitioners in 1988 showed that only 66 per cent of the respondents had been seen by a Visiting Practitioner in the previous two years; on average once per year in the South Island, and once every two years in the North Island (Scotney, 1990). Moreover, when asked about the desired frequency of visits from the Visiting Practitioners, 52 per cent stated that they wanted visits to remain as at present. Only 17 per cent requested more frequent visits.

The extent of coverage by the Visiting Pharmacist was even more limited. Only 54 per cent of general practitioners had been visited in the previous two years, and 66 per cent stated that they were satisfied with the present frequency of visits (Scotney, 1990). A 'small minority' of the respondents were critical of the services, stating that the information was either out of date by the time it was presented to the general practitioner or else the Visiting Professionals could not answer their queries.

The efforts of the Health Department provide a stark contrast to the resources deployed by the pharmaceutical industry, which in 1991 is training over 300 men and women throughout the country to become detailers (New Zealand Institute of Medical Representatives (NZIMR): 1991). According to a 1984 survey of general practitioners in the Whangarei region, 59.9 per cent were visited by detailers on three or more occasions per month, while only 8.9 per cent refused to see them (Benseman, 1985). No recent survey has been carried out to determine the frequency with which general practitioners are visited by 'drug reps'. However, from anecdotal reports, it appears that an increasingly popular form of contact with detailers consists of regular 'luncheon' meetings with a group of doctors from one practice. During such meetings, a promotional video is often shown. Lunch is, of course, provided by the company, and small gifts and free samples of the product may also be given (Durham, 1990). By co-ordinating the bookings, it is possible for a group of general practitioners to be entertained in this way by a different drug company every week.

The other form of prescribing information provided by the Health Department is the Clinical Notes Series. These currently comprise the *Clinical Services Letter* and *Therapeutic Notes*. Both newsletters are produced on an occasional basis and are widely distributed to all medical and dental practitioners, retail pharmacists, institutions such as the medical schools, and to interested individuals. The *Clinical Services Letter* provides information such as amendments to the Drug Tariff, changes to the schedule of medical fees and benefits, prescription guidelines, and the adverse medicines reaction reporting scheme. The *Therapeutic Notes* cover particular medical issues or conditions and usually present the opinions of invited contributors (Scotney, 1990).

A 1973 survey showed that 75 per cent of general practitioners read every issue of the *Therapeutic Notes* and that 88 per cent rated them good to excellent (Division of Clinical Services, 1980). These findings were supported by a 1984 poll of 45 general practitioners in the Whangarei area, in which 95 per cent of the respondents stated that they either read extensively or read and filed the mailings from the Health Department (Benseman, 1985). In the most recent national evaluation of the Clinical Notes Series conducted in 1988, just over half of the general practitioners stated that they always or sometimes read these newsletters fully. A further 40 per cent said that they scanned the newsletters for relevant information. The sections most referred to and regarded as the most useful were expert clinical commentary and schedules of fees and benefits. The most striking finding from this survey was that half of the respondents expressed a clear preference for more frequent and regular issues of the Series, rather than continuing to produce the newsletters on the present occasional basis.

In addition, 50 per cent of respondents also stated that they would like more drug cost and efficacy comparisons and more expert clinical comment (Scotney, 1990).

Clearly, general practitioners in New Zealand want more regular information about prescribing from non-commercial sources. The sparsity of such information provided by the Health Department again provides a stark contrast to the volume of promotional material mailed to doctors by pharmaceutical companies. For example, in 1989 the Department published two issues (each) of the *Therapeutics Notes* and *Clinical Services Letter*, totalling 30 pages; in contrast, one general practitioner reported having received 3.1 kilograms of mailed advertisements in the first four weeks of the same year (Durham, 1990).

The only other professional source of written information routinely available to doctors consists of scientific articles appearing in medical journals. Unfortunately, even peer-reviewed medical journals such as the *New Zealand Medical Journal* are not free of pharmaceutical advertising. Most journals rely heavily on advertising revenue to cover their expenses. A typical issue of the *New Zealand Medical Journal* during 1990 carried a full-page colour advertisement on every alternate page, rendering it impossible to read an article from beginning to end without casting a glance across at least one promotional message. More to the point, however, is the perceived lack of relevance of the articles carried in such journals for prescribing practice. For example, relatively few articles in the *New Zealand Medical Journal* are directly relevant to therapeutics. During 1989, ten clinical trials (six of them funded by drug companies) were published in a total of 21 issues of the Journal. Given the cost of subscribing to the Journal, it is perhaps not surprising that general practitioners look to alternative sources of prescribing information.

Commercial Sources of Prescribing Information

Pharmaceutical companies in New Zealand are estimated to spend approximately 13 per cent of their total sales dollar on advertising and promotion (Malcolm and Mehl, 1988). In 1989 to 1990 this would have amounted to $44.7 million, or the equivalent expenditure of $25,000 per general practitioner per year (Kawachi, 1990). Pharmaceutical promotion takes place in many forms. However, in terms of providing prescribing information to doctors, the most influential vehicles include the direct mailing of promotional material, the training and recruitment of detailers, and the financing of prescriber-oriented journals ('freebies' and 'throw-aways').

Other forms of pharmaceutical promotion exist, such as the distribution of small gifts and inducements (pens, diaries, table-mats, mugs, clocks, soaps, and medical equipment), competitions based on simple quizzes, with prizes such as cellular telephones, $2,000 holiday destinations, bottles of port (St George, 1989), and 'hospitality' provided in the form of free meals and alcohol, sponsored conference travel and accommodation, river rafting weekends and skiing trips (Kawachi, 1990). Such forms of promotion are deemed to have little educational content and hence will not be considered further. This does not imply, however, that they have *no* influence on doctors' prescribing habits. It would seem unlikely that a drug company would give away its shareholders' money in an act of disinterested generosity (Rawlins, 1984) — if gifts and inducements had no effect, market rationalism would have dispensed with such tactics long ago (Chren et al., 1989).

Drug Industry-financed Publications

Of the drug industry-financed journals ('freebies') distributed free of charge to medical practitioners in New Zealand, *New Ethicals* and *Patient Management* are probably the most widely read. Both journals are published monthly by ADIS Press Limited, an 'independent company not affiliated with any pharmaceutical manufacturer or association'. *New Ethicals*, subtitled 'New Zealand's journal of practical drug management', is circulated to 'selected medical practitioners and full-time chief hospital pharmacists', while *Patient Management*, subtitled 'the journal of practical patient care', is circulated to 'most general practitioners, house surgeons, selected registrars and specialists'.

Like the *New Zealand Medical Journal*, every alternate page in these 'freebies' carries a full-page drug advertisement. However, the similarity ends here, for unlike the *New Zealand Medical Journal*, these journals feature articles that are predominantly focused on practical drug management issues. They are also provided free of charge to all registered medical practitioners, and subscription can only be cancelled by failing to renew the annual practising licence, or by writing directly to the distributors. Most importantly, the articles in these 'freebies' are not usually peer-reviewed.

The lack of a refereeing process means that unless readers are familiar with the topic being dealt with, they have no way of judging the quality of the article. It might be scientifically valid or it might just reflect the author's individual opinion. To be fair, both the named 'freebies' have Honorary Editorial Boards made up of distinguished local pharmacologists and specialists in various fields, but it is not clear exactly how much input they have into the contents of the journals. As an example of editorial intrusion, the September 1990 issue of *New Ethicals* featured the summarized proceedings of a symposium on the management of hypertension held at the Otago Medical School. One of the papers presented at this symposium concerned the cost-effectiveness of treating mild hypertension with drugs (Kawachi and Malcolm, 1990). A summary of this study — which implied that drug treatment may not be cost-effective in certain types of patients — was reproduced in the Journal together with an editorial insertion stating:

There was vigorous debate following this presentation. Several speakers felt that many of the costs were estimates and therefore inaccurate, or believed they rested on invalid extrapolation from clinical trials.

The paper has since been published in an international peer-reviewed journal (Kawachi and Malcolm, 1991). Interestingly, this was the only paper in the entire summary of conference proceedings which received this type of editorial treatment. Papers presented by other authors at the same conference which supported the opposite view (i.e., the aggressive treatment of mild hypertension with drugs) were reproduced without editorial comment. The authors of the first paper were unable to protest this inconsistent treatment because, unlike most peer-reviewed journals, *New Ethicals* does not feature a correspondence column.

In the two surveys carried out to date on general practitioners' preferences for journals, the free journals were heavily favoured (West and Spears, 1976; Benseman, 1985). This result is not surprising. General practitioners have good reason to prefer this type of journal since its contents are usually more relevant to prescribing practice. The reliance of prescribers on commercial sources of information is a reflection of the failure of the Health Department and the medical profession to produce relevant, up-to-date, and peer-reviewed alternatives.

Probably the most common source of quick prescribing information for doctors is the *New Ethicals Catalogue*, also published by ADIS Press Limited. This soft-bound book (designed to fit snugly into the pockets of white coats) is distributed free of charge to all medical practitioners. Like its British counterpart, the *British National Formulary*, it contains a listing of all the medicines available in New Zealand, with their dosage, price, and Drug Tariff availability. However, unlike the *British National Formulary*, which refers to medicines by their generic names, medicines in the *New Ethicals Catalogue* are listed alphabetically according to brand name, thereby acting as a cue for New Zealand doctors to prescribe by brand name. At the very least, the *New Ethicals Catalogue* could be viewed as an advertisement to familiarize prescribers with individual brand names. In Canada, doctors also receive a catalogue called the *Compendium of Pharmaceuticals and Specialties* (CPS) which is partly financed by the drug advertisements it contains. A review of this publication concluded that it was basically a tool to promote the interests of the drug companies, and was neither a reliable nor an impartial reference (Bell and Osterman, 1983).

Direct Mailing of Promotional Material

In addition to the drug industry-financed journals and catalogues, medical practitioners receive a steady stream of mailed promotional material from individual companies. In one Christchurch practice, the seven members of the group received an average of six and a half pieces of unsolicited mail per week, or about 350 pieces per year (Isherwood et al., 1980). Based on the experiences of one general practice over a six-month period, it was estimated that drug companies in New Zealand currently spend over $590,000 per year in postage alone, sending 30,000 kilograms of advertisements printed on high quality paper to approximately 2300 general practitioners throughout the country (Clearwater, 1991). While there are occasional complaints from doctors about 'postal pollution' by the drug companies (Hay, 1974), most doctors apparently ignore the brochures and circulars they receive or throw them out after just glancing at them, even when they include a reprint from an academic journal (Benseman, 1985; Peacock, 1990). Partly in response to this sensitization of doctors, pharmaceutical companies have resorted to increasingly extreme tactics by which to gain their attention.

One Australasian public relations firm wrote in 1986:

> The new generation of doctors entering practice is more resistant to the traditional promotional tools, unreceptive to and sceptical of advertising claims, and in some cases unwilling to listen to what the company sales representative has to say . . . Consequently the industry is now recognising that the way to reach many of these doctors is through the 'softer', more objective approach that the public relations consultancy can provide. News releases and interviews with the medical and general media, seminars and special publications, and video and audio tapes are all means of educating, informing and at times even persuading general practitioners, often through the use of 'third party' endorsements by specialists or other opinion leaders (Network Communications, 1986).

The following example illustrates the 'softer, more objective' approach to pharmaceutical promotion. During 1990, the manufacturers of Capoten [captopril-Squibb] ran a campaign in which they bought time on national television to make a public broadcast. The video featured a panel of local specialists being interviewed by former network newsreader Dougal Stevenson on 'an important breakthrough in our understanding of hypertension'. The video summarized the findings of a clinical trial by Pollare et al. (1989) — funded by Squibb and published in the *New England Journal of Medicine* in 1989 — which compared the metabolic effects of captopril against an alternative antihypertensive drug, hydrochlorothiazide.

Interestingly, the promotional activities surrounding the same study were singled out for criticism in a subsequent editorial which featured in the *Journal of the American Medical Association*:

> The widespread publicity and public response to a recent study by Pollare et al. . . . serve to highlight the impact of the media and pharmaceutical industry on the practice of medicine. Patients' reactions were predictable following a front-page story in the *New York Times*, articles in more than 200 newspapers, and national television coverage. Many patients expressed concerns about their treatment despite the fact that their hypertension may have been well controlled by diuretic therapy (of which hydrochlorothiazide is one) for many years . . . The production of 30-, 60- and 90-second audio and video spots, which were sent to radio and television stations across the country, followed a press conference called by a major pharmaceutical company . . . The conference was put together by a public relations firm (at present the Food and Drug Administration has no control over the content of such conferences) . . . Some physicians who may have been influenced by the reports of the press conference or were not aware that the data were not entirely new or decisive may have switched patients to other drugs. Based on our personal observations, many physicians have indeed been persuaded by both the campaign by the lay media and an extensive medical promotional campaign, which included a national cable television program for physicians, to change treatment practice (Moser et al., 1991).

The promotional video shown on New Zealand television ended with an announcement calling for entries to a competition for viewers, the main prize being a television and video player. Following this broadcast a copy of the video was distributed to all medical practitioners in the country.

Both the Medicines Regulations (1984) and the voluntary code of advertising practice of the Researched Medicines Industry (RMI) require a minimum set of information to be printed on promotional material, which includes headings such as 'indications for use', 'contraindications', and 'adverse reactions'. Yet these requirements have been shown to be frequently breached (Lexchin, 1989; Peacock, 1990). Neither the Health Department nor the RMI monitor the frequency with which their own regulations and codes are breached. The Health Department has admitted to lacking the necessary resources and personnel to enforce the regulations controlling promotion (Boyd, 1990), while the RMI will institute an investigation only if a complaint is received, usually from a rival company adversely affected by the infringing advertisement. An unknown proportion of complaints are handled informally and never reach the RMI's Disciplinary Committee. Even where a company is found to have breached the code, neither the ruling nor the sanctions applied are ever made public. According to Bill McLauchlan, former Chief Executive Officer of the RMI, such information is 'probably not of great concern to the consumer' (Consumers Institute, 1990).

A number of doctors firmly believe that members of the medical profession are immune to promotional influence. Thus a Professor of the Department of General Practice at the University of Otago wrote (in a guest editorial in a 'freebie' diary given away to medical practitioners):

On behalf of my colleagues, I resent most bitterly the assertions made by academics which infer to the public at large that working general practitioners are . . . seducible, manipulated, needing advice from the Health Department, not trained in clinical evaluation and even lacking in eyesight where the industry puts the side effects of pills in small print — in other words innocent but stupid (Murdoch, 1990a).

The facts, however, speak for themselves. Studies have consistently found that the more intensively a product is advertised in journals the greater its market share (Montgomery and Silk, 1972; Leffler, 1981; Krupka and Vener, 1985) and that the more physicians remember a drug advertisement the more likely they are to prescribe the product (Walton, 1980; Healthcare Communications, 1989). These findings do not imply that general practitioners are 'stupid'; rather, they tend to suggest that when it comes to being open to the powers of suggestion, doctors respond no differently from any other member of the general public. What no doubt *does* surprise many viewers outside the profession is what Michael Rawlins described as 'the [extraordinary] degree to which the profession, mainly composed of honourable and decent people, can practise such self-deceit' when it comes to denying *any* effect of pharmaceutical advertising on prescribing (Rawlins, 1984).

Detailers

Pharmaceutical representatives (or 'detailers') are employees of drug companies who are paid to visit surgeries in order to promote the products of their company. They are among the most influential sources of prescribing information. In a survey of 332 general practitioners in England, pharmaceutical company representatives were identified as the 'most useful' commercial source of information out of a list of 18 sources. Consistent with previous studies, an overwhelming proportion of general practitioners considered themselves to be influenced by professional rather than commercial sources of prescribing information. Nonetheless, 36 per cent still considered the 'drug rep' to be a 'very important' influence upon their prescribing behaviour. Moreover, the representative was mentioned as the source of knowledge for the last product prescribed in 58 per cent of cases, over four times the prevalence of mention of any other information source (Greenwood, 1989a). In the case of one product, some 77 per cent of general practitioners favoured the 'commercial' view of the product rather than the weight of medical evidence, while 'commercial' views of three other products were favoured by 55 per cent, 28 per cent and 13 per cent of doctors respectively (Greenwood, 1989b).

The results of this survey confirmed what the pharmaceutical industry has known for a long time. As one marketing director of a large company expressed it to a conference of pharmaceutical marketing personnel in the United Kingdom:

The way the company gets through to the doctor is through its best communicator — the representative. Furthermore, the Marketing Director knows well — in spite of popular comment to the contrary — that a large percentage of doctors rely on pharmaceutical representatives for product information, and these doctors at the time of the interview can be influenced (quoted in Greenwood, 1989a).

The available evidence suggests that New Zealand prescribers see company representatives just as frequently as their overseas counterparts. Although some New Zealand doctors feel that the quality of service from detailers is highly variable the majority consider them a good source of information (Martin, 1977; McMillan, 1982; Benseman, 1985). The drug industry in New Zealand devotes approximately 60 per cent of its promotional budget to detailers (Lexchin, 1989). However, nobody has really looked at what New Zealand detailers do. A survey of representatives from 21 firms throughout the United Kingdom provides some clues to this question. Apart from promoting the company's products, sales representatives routinely gather market intelligence. This consists of monitoring the prescribing behaviour of individual general practitioners, charting the movement of pharmaceuticals through wholesalers and retail outlets, and supplementing this information through contact with pharmacists (Greenwood, 1989a).

Pharmaceutical representatives are thus employed to fulfil a dual function — as company salespersons on the one hand (which includes tasks such as gathering market intelligence) — and as 'detailers' delivering prescribing information on the other. Which role predominates in their professional activities may be a matter of debate. Nevertheless, sooner or later an unavoidable conflict must arise between the sales motive and the objective of providing unbiased prescribing information. Recognizing this conflict of interest, the World Health Organization (1988) has explicitly recommended that representatives 'should possess sufficient medical and technical knowledge and integrity to present information on products', and furthermore that 'the main part of the remuneration of medical representatives should not be directly related to the volume of sales they generate'. In this respect, the findings of previous studies give little cause for reassurance. In the United Kingdom survey of sales 'reps' mentioned above, the average training courses were about five weeks in duration. Some 86 per cent of representatives were found to be given sales targets to achieve — a feature of their jobs which appears to conflict with the principles of rational prescribing (Greenwood, 1989a). An American detailer was even more frank in describing his job as that of convincing doctors to use his company's product in preference to the competitor's 'by any means I could'; while according to a Canadian detailer his job entailed 'perceiving needs, or creating them' (quoted in Lexchin, 1989, p. 666). A Finnish study indicated that detailers seldom tended to mention side-effects of drugs and contraindications to using the drugs they were promoting (Hemminki, 1977).

In New Zealand, company representatives are trained at the NZIMR, a drug industry-financed institution which offers a two-year correspondence course. Whether this training is adequate to equip detailers with 'sufficient medical and technical knowledge' is unknown; no thorough assessment of the Institute or its graduates has yet been made (Lexchin, 1988).

Doctors and the Pharmaceutical Industry — Too Close for Comfort?

The drug industry has a perfect right to promote its products. Indeed, pharmaceutical promotion has important benefits. Doctors need information about drugs, especially the existence of new products. Insofar as advertising provides this information and helps doctors to choose the most appropriate drug for their patients, then it is desirable. In the absence of satisfactory state-produced alternatives, doctors rely heavily on commercially-produced catalogues, mailed advertisements and detailers for their prescribing information, particularly in relation to new products.

A substantial portion of postgraduate medical education in New Zealand is further financed by pharmaceutical companies, operating via sponsorship of conferences, symposia, and seminars (Lexchin, 1988; Durham, 1990). In a survey of 284 doctors, 70 per cent rated company-sponsored seminars as one of their three most important sources of information (Martin, 1977). Often these seminars are accompanied by entertainment for the participants, providing an extra incentive for doctors to take part in continuing education. The Obstetrical and Gynaecological Society Conference in Auckland in February 1987 included overseas speakers funded by Pharmaco and Ciba-Geigy, wine and cheese courtesy of Douglas Pharmaceuticals, conference and dinner entertainment sponsored by Schering, a Waitemata harbour cruise subsidized by Allen and Hanbury, and a cocktail hour in the Regatta Room courtesy of Ciba-Geigy (Coney, 1987). In the last two years, Wellington general practitioners have been invited to weekend meetings in Rotorua, the Marlborough Sounds, Akaroa, and Queenstown. They were also invited on a rafting expedition on the Tongariro River, for which the company paid all the costs of the weekend, including transport from Wellington; in return, the total educational content of the programme was one three-hour discussion of a cardiovascular topic (Durham, 1990).

Returning to the previous example of the captopril promotion in the United States, it was reported that:

Numerous symposia have been held throughout the country reiterating the findings of the study by Pollare et al. (1989) . . . These meetings, as well as other meetings in other areas of medicine, have been advertised as educational events. They are, however, as many physicians have come to realize, largely promotional. No one can object to an industry-sponsored meeting to launch a new drug. Physicians who accept an invitation to attend know what to expect — but the type of 1- and 2-day promotional symposia that are presently proliferating under the guise of updates or new approaches should be discouraged. We must question why this type of industry pressure is not challenged more vigorously by major medical organizations . . . If we are to continue to practise medicine based on scientific data and its appropriate dissemination by medical journals and not industry-sponsored promotional medical education programs and/or public relations-inspired media efforts, we must take a stand against the type of prescribing pressure that was applied following the publication of the study by Pollare et al. — and indeed studies relating to other drugs. Otherwise, we might just as well give our prescription pads to our media and industry colleagues (Moser et al., 1991).

Pharmaceutical companies appear to be spending more of their promotional budget on funding doctors' educational meetings. In the United States, the Senate Labor and Human Resources Committee (chaired by Senator

Edward Kennedy) recently reported that in 1988, eighteen major pharmaceutical companies spent a total of $86 million on funding symposia, a fourteen-fold increase from the $6 million (inflation-adjusted) spent by the same companies in 1975 (Randall, 1991a).

In 1984 Rawlins noted that:

> Doctors are becoming so accustomed to sponsored postgraduate medical education that it is difficult to attract them to meetings where they have to pay for their own registration and refreshments. Postgraduate education is thus tending to become the responsibility of the drug industry rather than of postgraduate deans, clinical tutors, and the profession itself. This trend should be a major concern to us all, because of the potential for distorting postgraduate medical education away from the needs of patients and the health service, towards the requirements of the industry (Rawlins, 1984).

The dilemma faced by prescribers in today's market is one of finding the optimal balance between reliance on commercial as opposed to non-commercial sources of information. This task is made all the more difficult by the fact that the vast majority of the drugs currently in use have been developed in the last 10 or 15 years. One study showed that approximately 85 per cent of all prescriptions written by senior physicians who graduated from medical school in 1960 will be for a drug about which they have received no formal education (American College of Physicians, 1988). On the other hand, only about 3 per cent of new drugs are deemed to represent an important therapeutic advance (Randall, 1991a). Doctors are likely to be able to obtain information about this small fraction of drugs from sources other than the pharmaceutical industry. Furthermore, prescribers do not actually need to learn about hundreds of drugs. For example, over 50 per cent of all prescriptions written by family practitioners in Canada are for no more than 27 different medicines (Lexchin, 1990). Hence it is not difficult for prescribers to consult objective sources to obtain information about the small number of drugs that they actually use, rather than to rely heavily on commercial sources. Whilst acknowledging the need for pharmaceutical companies to promote their products, both doctors and state regulators thus need to recognize the hazards of undue influence.

In September 1990 the New Zealand Medical Association (NZMA) issued a set of position statements concerning the relationship between the medical profession and the pharmaceutical industry (NZMA, 1990). Like its counterparts issued by medical organizations in other parts of the world, the guidelines formulated by the NZMA are tentative, and (to borrow Richard Smith's assessment of the Royal College of Physicians' Code), 'castrated by the impossibility of definition' (Smith, 1986). For example, the NZMA Code recommends that 'professional societies should develop . . . guidelines to discourage excessive industry-sponsored gifts, amenities, and hospitality to doctors at meetings', without actually defining what is meant by the term 'excessive'.

When it comes to defining what is 'acceptable' in terms of perks and kickbacks, studies have indicated that doctors tend to apply softer ethical values within their own profession. In one United States study medical students taking part in a seminar on ethics were asked to appraise the acceptability of two different scenarios: in the first scenario an elected public official who awards contracts accepts $50 from a prospective bidder; while in the other scenario a medical student who will eventually prescribe drugs accepts a $50 gift from a drug company. While 85.4 per cent of the students thought it was improper for the public official to accept a $50 gift, only 46 per cent believed it was improper for a medical student to accept such a gift from a drug company (Palmisano and Edelstein, 1980).

In contrast to these values held by the medical profession, the corporate policies of most pharmaceutical companies generally forbid individual employees from receiving gifts from outside suppliers or customers. Inquiries made at major companies such as Merck Sharp and Dohme, Upjohn, Eli Lilly and Company, and Abbott Laboratories reportedly revealed that all have corporate-wide policies that either forbid the acceptance of gifts or limit the value of gifts to five or ten dollars. According to one manager of a major drug company, her staff had returned numerous gifts above a nominal value, including 'a gorgeous wooden golf set and a case of beer'; she added 'we want to keep things straight, aboveboard, business' (Randall, 1991b).

Does the relationship between the medical profession and the pharmaceutical industry pose a threat to rational prescribing? There are conflicting answers to this question, and the debate continues in the medical press (Lexchin, 1987; Hardy, 1987, Thompson, 1990; MacLauchlan, 1990; Murdoch, 1990a; Murdoch, 1990b; Reid, 1990). To those who answer in the affirmative, the current relationship between doctors and the drug industry threatens to undermine the trust invested by the public in prescribers as unbiased agents acting in the best interests of patients and taxpayers. In the words of Senator Edward Kennedy who chaired the 1990 Senate Labor and Human Resources Committee inquiry into drug promotion: 'Doctors who accept lavish industry gifts are jeopardizing their objectivity and compromising the trust of their patients' (Randall, 1991a). To others who deny any undue influence, the risk is merely theoretical; even the hint of a suggestion that the promotional activities of drug companies might influence prescribing is, in the words of an RMI press release (4 May 1990): 'a prejudice implied against the ethics and intelligence of working doctors'. However, according to the findings of a classic study by Avorn:

Although the vast majority of practitioners perceived themselves as paying little attention to drug advertisements and detail men, as compared with papers in the scientific literature, their beliefs about the effectiveness of the [studied] drugs revealed quite the opposite pattern of influence in large segments of the sample . . . Rather than coming as a surprise, the predominance of nonscientific rather than scientific sources of drug information is consistent with what would be predicted from communications theory and marketing research data. Drug advertisements are simply more visually arresting and conceptually accessible than are papers in the medical literature, and physicians appear to respond to this difference (Avorn et al., 1982).

Wherever the ultimate truth may lie in this continuum of opinion, the growing disquiet about the relationship between the profession and the industry is an international phenomenon. Public relations firms have become alerted to this trend and in 1986 gave the following advice to the industry:

In both Australia and New Zealand, the pharmaceutical industry is currently facing the challenges of an increasingly sophisticated consumer movement, and industry executives are relying on communications initiatives to ensure their side of the story is presented effectively to the public . . . Events in recent months in both Australia and New Zealand have highlighted the potential threat which consumerism poses to the pharmaceutical industry . . . Several recent health-related issues affecting the pharmaceutical industry in Australasia have been part of a concerted international campaign by consumer groups. A decade, or even five years ago, such issues would not have gone beyond Europe and the USA. Nowadays, anywhere in the world is vulnerable (Network Communications, 1986).

Amidst the growing 'threat' of consumerism, the industry has increasingly sought to align itself with doctors and the lay public. Thus a press release issued by the RMI warned that:

Attacks . . . are [being] made on primary health care providers (including doctors and specialists) who attempt to protect their prerogative to prescribe the medicine they consider best for their patients (Researched Medicines Industry, 1990a).

In a pamphlet distributed to the lay public, the RMI similarly advised that:

Politicians and Civil Servants . . . want to spend less of your tax on health . . . they're looking in the bargain basement. Buying cheaper, less effective drugs, rather than the medicines that are best for you . . . Your family Doctor knows your medical problems personally . . . But with the political penny-pinching, your doctor may be having to make decisions based on what he or she believes you can afford . . . This is unfair to both you and your doctor. Doctors are committed to providing the best treatment and should not be forced to choose between your needs and Government cost-cutting . . . [Therefore] ask your doctor and pharmacist about new medicines for your illness. Discuss all the alternatives with your doctor and consider whether cost is limiting your options. Exercise your democratic right (Researched Medicines Industry, 1990b).

By conjuring up the bugbear that administrators and civil servants may eventually usurp doctors' clinical freedom to prescribe whatever is best for their patients, the medical profession is thus becoming gradually sensitized to vote against any moves to regulate pharmaceutical promotion.

Options for Action

In a recent review of business regulatory agencies in Australia, it was concluded that while the regulatory coverage in the pharmaceutical industry was comprehensive, the resources deployed to undertake enforcement tasks were slight. In fact, over 70 per cent of the Department of Health officers concerned with drug regulation were employed in the routine task of processing Pharmaceutical Benefits Scheme claims (Grabosky and Braithwaite, 1985). A similar conclusion was drawn in an external review of the Medicines and Benefits section of the New Zealand Health Department (Coopers and Lybrand Associates, 1986).

According to Grabosky and Braithwaite, the benign nature of business regulation in Australia did not appear to be determined by any inadequacy of powers at the disposal of regulatory agencies. The Minister of Health in New Zealand is similarly vested with considerable powers within the Medicines Act (1981) and the Medicines Regulations (1984) to enforce the legislation concerning pharmaceutical promotion. Instead, the conduct of regulatory agencies in both countries appears to be characterized by an overriding concern to avoid any confrontation with the industry. In Grabosky and Braithwaite's analysis, the actions of regulatory agencies consisted mainly of:

Platitudinous appeals to industry to act responsibly; token enforcement targeted in a manner which bore no necessary relationship to failures to heed those platitudinous appeals; keeping the lid on problems which could blow up into scandals; and passing the buck to another agency when the lid could not be kept on a scandal (Grabosky and Braithwaite, 1985).

In the area of pharmaceutical promotion, *laissez-faire* passes for Government policy. For the moment, the buck seems to have been passed to the industry to regulate itself. Whether the industry and the medical profession can heal themselves, or whether additional regulation or legislation is needed, is a matter for debate. More regulation, as recently advocated by the Public Health Association working party (Kawachi, 1990), may not necessarily yield the most satisfactory solution. Indeed, greater state intervention may lead, as some have warned, to 'significant hidden costs, frustrated prescribers, a switched-off industry and a nonprogressive, inwardly-thinking pharmaceutical management' (Maling, 1989). In some countries, for example Sweden, limited successes have been claimed for models of industrial self-regulation (Perman, 1990). However, these are exceptions rather than the rule; in most countries, regulatory approaches have so far failed to curb promotional excesses (Herxheimer and Collier, 1990).

An emerging set of strategies emphasizes the education of prescribers through non-commercial sources of information ('counter-promotional education'). For some time in the United Kingdom non-commercial prescribing information has been provided free of charge to doctors in the form of the *British National Formulary* (a compendium of drugs referenced according to generic names, published jointly by the British Medical Association and the Royal Pharmaceutical Society of Great Britain); the *Drug and Therapeutics Bulletin* (a fortnightly publication in the form of a newsletter, often featuring prescribing guidelines, prepared by the Consumers Association); and data sheets containing detailed information about new drugs, which the British pharmaceutical companies are legally obliged to distribute to prescribers. There are at present no equivalent forms of any of these information sources available to New Zealand prescribers. Based on the findings from surveys of general practitioners' information sources (Benseman, 1985; Scotney, 1990), there would appear to be a strong case for the Department of Health to finance similar publications — or at least to require the drug industry to provide practitioners with data sheets for new drugs.

In 1989 a new project was set up in the Nelson area to promote cost-effective prescribing in general practice. The Nelson General Practice Prescribing Project is a joint initiative between the Nelson sub-faculty of the Royal New Zealand College of General Practitioners, the Wellington School of Medicine, the Department of Health, and the Nelson-Marlborough Area Health Board. The project has so far involved the development of a local Preferred Medicines List; the provision of individualized prescription audit reports to participating general practitioners; the use of one-to-one educational outreach visits by a trained pharmacist; and the development of a targeted educational package. To date the project has found high acceptability among general practitioners, and early reports indicate improvement in cost-effective prescribing (Ferguson and Maling, 1990).

In Britain a pilot programme is under way to provide general practitioners within the Trent Regional Health Authority with access to 'state detailers' (Greenwood, 1989a, 1989b). The state detailers are recruited by the Regional Health Authority to provide independent, professional prescribing information to general practitioners. In a critical analysis of all the existing interventions employed to improve drug prescribing, Soumerai et al. (1989) identified the use of one-to-one educational outreach visits by specially-trained clinical pharmacists or physician counsellors as one of the most effective strategies by which to modify and improve the prescribing behaviour of primary care physicians. The pharmaceutical industry has known this for some time through the use of company representatives. However, the time has probably come for the state to recruit its own detailers in order to provide information to balance the claims made by commercial interests. In New Zealand, the present Visiting Professionals scheme is too woefully under-resourced to perform this task.

Finally, providing doctors with better information is only a first step to better prescribing — doctors also have to use that information. There are some models in New Zealand which show how rational prescribing can be achieved. Eight doctors at the Christchurch South Health Centre set aside one and a half hours each week for case presentations, patient management discussions, and professional education (Isherwood et al., 1980; 1982). Although the peer review system has its problems, it also has potential to monitor and improve prescribing behaviour (Durham, 1988; Soumerai et al., 1989). In the Christchurch model practice the drugs prescribed were shown to be related rationally to patient diagnosis: patients with more serious and chronic conditions received more medicines per script; and patients with psychological/psychiatric or social problems tended to receive longer consultations and were more likely to receive counselling from their doctors. Doctors at this Health Centre also wrote fewer prescriptions per patient compared to the Christchurch District overall, and their prescriptions were significantly less costly (Isherwood et al., 1980; 1982).

Most general practitioners in New Zealand are paid on a fee-for-service basis. There are other ways of remunerating doctors: they can be salaried, or they can be paid under a capitation system. In the latter arrangement, patients formally register with practitioners, and doctors then receive a set sum of money per patient per month regardless of whether the patient actually visits them. With both of these arrangements, there is no financial incentive to see a large number of patients per day, and doctors have more time to spend with patients. Canadian experience has shown that salaried doctors working in community centres are better prescribers than those on a fee-for-service system (Renaud et al., 1980).

This chapter concludes with a sobering tale from America. In 1987 a company called Genentech released a new clot-dissolving drug onto the United States market. The drug, called TPA (tissue plasminogen activator) cost $US2,500 per dose compared to the existing alternative, streptokinase, manufactured by Astra and costing only $US220. Amidst a 'blinding flash of pitchmen, promotion and public relations hoo-ha', TPA quickly gained the dominant market share over its rival, selling over $US600 million each year worldwide (Purvis, 1991). Then in March 1991, the findings of the International Study of Infarct Survival (ISIS-3) were revealed at a meeting of American cardiologists in Atlanta. The study, involving 46,000 patients in over 1000 hospitals worldwide, was a clinical trial comparing the benefits and risks of streptokinase versus TPA. The results demonstrated that both drugs were equally effective in dissolving clots and keeping people alive. However, the drugs differed in their side-effects, with 1.5 per cent of TPA-treated patients suffering a stroke, compared to 1.1 per cent among streptokinase-treated patients. According to reports, Genentech, the manufacturer of TPA 'was underwhelmed' (Anonymous, 1991). At a press conference, the company maintained that 'most US cardiologists will view the results as inconclusive for US medical practice'. Apparently, Genentech's predictions were correct.

Astra, the manufacturer of streptokinase presented a poster at the Atlanta meeting crowing that their product had at last proven to be as effective as TPA: 'But this hard-won truth has had little impact on the company's share prices.' According to a report filed by the *Economist*: 'That is apparently because it will make little difference to the drugs' success on the market. Pharmaceutical analysts say they no longer bother with considering such studies. It is the marketing that counts' (Anonymous, 1991).

References

American College of Physicians, Health and Public Policy Committee (1988) Improving medical education in therapeutics *Annals of Internal Medicine* 108: 145–7.

Anonymous (16 March 1991) Heart-attack drugs. Trials and tribulations *The Economist*: 88–9.

Avorn, J., Chen, M., and Hartley, R. (1982) Scientific versus commercial sources of influence on the prescribing behavior of physicians *American Journal of Medicine* 73: 4–8.

Becker, M., Stolley, P., Lasagna, L., McEvilla, J., and Sloane, L. (1972) Differential education concerning therapeutics and resultant physician prescribing patterns *Journal of Medical Education* 47: 118–27.

Bell, R. and Osterman, J. (1983) The Compendium of Pharmaceuticals and Specialties: a critical analysis *International Journal of Health Services* 13: 107–18.

Benseman, J. (1985) The great paper waste: the use of unsolicited medical literature by general practitioners *New Zealand Family Physician* 12: 96–8.

Blondeel, L., Cannoodt, L., De Meyere, M., and Proesmans, H. (1987) *Prescription behaviour of 358 Flemish general practitioners* Prague, International Society of General Medicine meeting.

Boyd, B. (1990) What you see depends on where you sit. In Kawachi, I. (ed) *Pharmaceutical Advertising and Promotion — Options for Action* Wellington, Public Health Association of New Zealand 50–1.

Chren, M., Landefeld, S., and Murray, T.H. (1989) Doctors, drug companies, and gifts *Journal of the American Medical Association* 262: 3448–51.

Clearwater, G. (1991) Advertising junk mail *New Zealand Medical Journal* 104: 344.

Coney, S. (11 October 1987) We're ripe for a drugs bust *Dominion Sunday Times* Wellington.

Consumers Institute (April 1990) Undue Influence. How drug companies sway doctors' decisions *Consumer Food and Health* 7–9.

Coopers and Lybrand Associates (1986) *Removal of Medicines from Price Control* Wellington, Department of Health.

Division of Clinical Services (1980) *Medicine Control in New Zealand* Wellington, Government Printer.

Durham, J. (1988) Prescribing information and peer review in general practice. In Maling, T.J.B. (ed) *Drug Prescribing Efficiency* Auckland, Medical Publishing Company Limited 42–4.

Durham, J. (1990) Pharmaceutical promotion in New Zealand — a view from general practice. In Kawachi, I. (ed) *Pharmaceutical Advertising and Promotion — Options for Action* Wellington, Public Health Association of New Zealand 43–9.

Ferguson, R. and Maling, T. (1990) *The Nelson General Practice Prescribing Project* Phase 1 Project Report. Wellington, Wellington School of Medicine.

Ferry, M., Lamy, P., and Becker, L. (1985) Physicians' knowledge of prescribing for the elderly. A study of primary care physicians in Pennsylvania *Journal of the American Geriatric Society* 33: 616–21.

Grabosky, P. and Braithwaite, J. (1985) *Of Manners Gentle. Enforcement strategies of Australian Business Regulatory Agencies* Melbourne, Oxford University Press.

Greenwood, J. (August 1989a) Prescribing and salesmanship *HAI (Health Action International) News* No. 48: 1–2, 11.

Greenwood, J. (1989b) *Pharmaceutical Representatives and the Prescribing of Drugs by Family Doctors* (PhD thesis abstract): 11. Department of Administrative and Social Studies, Teesside Polytechnic, Middlesborough, Cleveland, United Kingdom.

Haayer, F. (1982) Rational prescribing and sources of information *Social Science and Medicine* 16: 2017–23.

Hardy, A.J. (1987) Drug promotion *New Zealand Medical Journal* 100: 667.

Hay, D. (1974) Postal pollution *New Zealand Medical Journal* 80: 464.

Healthcare Communications I (1989) *The Effect of Journal Advertising on Market Shares of New Prescriptions* New York, The Association of Independent Medical Publications Inc.

Hemminki, E. (1977) Content analysis of drug-detailing by pharmaceutical representatives *Medical Education* 11: 210–15.

Herxheimer, A. and Collier, J. (1990) Promotion by the British pharmaceutical industry, 1983–8: a critical analysis of self regulation *British Medical Journal* 300: 307–11.

Isherwood, J., Malcolm, L.A., and Hornblow, A. (1980) *Factors in general practitioner prescribing* Christchurch, Health Planning and Research Unit.

Isherwood, J., Malcolm, L.A., and Hornblow, A. (1982) Factors associated with variations in general practitioner prescribing costs *New Zealand Medical Journal* 95: 14–17.

Kawachi, I. (ed) (1990) *Pharmaceutical Advertising and Promotion — Options for Action* Wellington, Public Health Association of New Zealand.

Kawachi, I. and Malcolm, L.A. (September 1990) Cost-effectiveness in treating hypertension. Summary of a paper presented at the Satellite Symposium of the XIth World Congress of Cardiology, Dunedin, 5–7 February 1990 *New Ethicals* 27: 38–44.

Kawachi, I. and Malcolm, L.A. (1991) The cost-effectiveness of treating mild to moderate hypertension — a reappraisal *Journal of Hypertension* 9: 199–208.

Krupka, L. and Vener, A. (1985) Prescription drug advertising: trends and implications *Social Science and Medicine* 20: 191–7.

Leffler, K. (1981) Persuasion or information? The economics of prescription drug advertisement *Journal of Law and Economics* 24: 45–74.

Lexchin, J. (1987) Drug promotion *New Zealand Medical Journal* 100: 603.

Lexchin, J. (1988) Pharmaceutical promotion in New Zealand *Community Health Studies* 12: 264–72.

Lexchin, J. (1989) Doctors and detailers: therapeutic education or pharmaceutical promotion? *International Journal of Health Services* 19: 663–79.

Lexchin, J. (1990) Prescribing by Canadian general practitioners: a review of the English language literature *Canadian Family Physician* 36: 465–70.

Linn, L. and Davis, M. (1972) Physicians' orientation toward the legitimacy of drug use and their preferred source of new drug information *Social Science and Medicine* 6: 199–203.

MacLauchlan, B. (1990) For the price of a pen! *New Zealand Medical Journal* 103: 250.

Malcolm, L.A. and Mehl, A. (1988) Recent trends in drug prescribing rates and costs in New Zealand *New Zealand Medical Journal* 101: 233–6.

Maling, T.J.B. (1989) The prescribing process and options for modification *New Zealand Medical Journal* 102: 43–5.

Mapes, R. (1977) Aspects of British general practitioners' prescribing *Medical Care* 15: 371–81.

Martin, R. (1977) Drug costs *New Zealand Medical Journal* 86: 202.

McMillan, J. (1982) Information about medicines *New Zealand Medical Journal* 95: 674–5.

Montgomery, D. and Silk, A. (1972) Estimating dynamic effects of market communications expenditures *Management Science* 18: B485–B501.

Moser, M., Blaufox, M.D., Freis, E., Gifford, R.W. Jr., Kirkendall, W., Langford, H., Shapiro, A., and Sheps, S. (1991) Who really determines your patients' prescriptions? *Journal of the American Medical Association* 265: 498–500.

Murdoch, J.C. (August 1990a) *The profession and the research-based pharmaceutical industry — time for a reappraisal* (guest editorial), Diary of Events for the medical and health professions. Palmerston North, Allen and Hanburys 1–5.

Murdoch, J.C. (1990b) For the price of a pen! *New Zealand Medical Journal* 103: 190.

Network Communications Public Relations Consultants (November 1986) PR: a new force in the marketing mix *Communicator*.

New Zealand Medical Association (NZMA) (13 September 1990) *Position statements on the relation between the medical profession and the pharmaceutical industry* Wellington.

Palmisano, P. and Edelstein, J. (1980) Teaching drug promotion abuses to health profession students *Journal of Medical Education* 55: 453–5.

Peacock, S. (1990) Two surveys of pharmaceutical promotion and advertising. In Kawachi, I. (ed) *Pharmaceutical Advertising and Promotion — Options for Action* Wellington, Public Health Association of New Zealand 54–6.

Peay, M.Y. and Peay, E.R. (1988) The role of commercial sources in the adoption of a new drug *Social Science and Medicine* 12: 1183–9.

Perman, E. (1990) Voluntary control of drug promotion in Sweden *New England Journal of Medicine* 323: 616–7.

Pollare, T., Lithell, H., and Erne, C. (1989) A comparison of the effects of hydrochlorothiazide and captopril on glucose and lipid metabolism in patients with hypertension *New England Journal of Medicine* 321: 868–73.

Purvis, A. (18 March 1991) Cheaper can be better *Time Magazine* 60.

Randall, T. (1991a) Kennedy Hearings say no more free lunch — or much else — from drug firms *Journal of the American Medical Association* 265: 440–2.

Randall, T. (1991b) Ethics of receiving gifts considered *Journal of the American Medical Association* 265: 442–3.

Randall, T. (1991c) Does advertising influence physicians? *Journal of the American Medical Association* 265: 443.

Rawlins, M.D. (1984) Doctors and the drug makers *Lancet* ii: 276–8.

Reid, B. (1990) PHA doesn't live in 'real world' *New Zealand General Practice* 1–2 ADIS Press Limited.

Renaud, M., Beauchemin, J., Lalonde, C., Poirier, H., and Berthiaume, S. (1980) Practice settings and prescribing profiles: the simulation of tension headaches to general practitioners working in different practice settings in the Montreal area *American Journal of Public Health* 70: 1068–73.

Researched Medicines Industry (RMI) (4 May 1990a) *Press release* Wellington.

Researched Medicines Industry (RMI) (1990b) *Your right to better medicines* (pamphlet) Wellington, New Zealand.

Scotney, T. (1990) Information services to general practitioners: a review of the Clinical Services Letter, Therapeutic Notes and the Visiting Professionals *New Zealand Family Physician* 17: 9–11.

Simpson, J., Squires, I., and Elliott, J. (1984) General practice caseload management *New Zealand Family Physician* 11: 67–71.

Smith, R. (1986) Doctors and the drug industry: too close for comfort *British Medical Journal* 293: 905-6.

Soumerai, S., McLaughlin, T., and Avorn, J. (1989) Improving drug prescribing in primary care: a critical analysis of the experimental literature *Milbank Quarterly* 67: 268-317.

St George, I. (1989) The competitions (editorial) *New Zealand Family Physician* 16: 186-7.

Thompson, A. (1970) Controlling the costs of community medical care. In Dodge, J. (ed) *The organization and evaluation of medical care* Dunedin, University of Otago 199-221.

Thomson, A. (1990) For the price of a pen! *New Zealand Medical Journal* 103: 135.

Walton, H. (20 June 1980) Ad recognition and prescribing by physicians *Journal of Advertising Research* 39-48.

West, S.R. and Spears, G. (1976) The continuing education needs of general practitioners *New Zealand Medical Journal* 84: 225-9.

Wolfe, S.M. (June 1991) Promotional practices in the pharmaceutical industry *Health Action International News* No. 59: 1-2, 10-11.

World Health Organization (1988) *Ethical criteria for medicinal drug promotion* Geneva, World Health Organization.

Where's the Bite? — Six Case Studies of the Voluntary Regulation of Pharmaceutical Advertising and Promotion

Ichiro Kawachi

Introduction

In developed countries, Government regulation of the pharmaceutical industry encompasses a wide range of activities including the monitoring of product development, approval of products for marketing, regulation of prices, and the monitoring of adverse drug reactions. The regulation of pharmaceutical advertising and promotion is, however, one area which appears to be under-resourced by governments in many countries. Although virtually all developed countries have laws and regulations governing the ethical marketing of medicinal products, few governments appear willing to take up the dominant role in regulating the marketing practices of the pharmaceutical industry. Regulation has been left instead to self-policing by the industry and the medical profession.

The first part of this chapter will review the successes and failures of voluntary regulation of pharmaceutical promotion in six different countries: Britain, the United States, Canada, Australia, Sweden, and New Zealand. The second part outlines some general recommendations for further improvements, based upon the lessons learned from abroad.

Regulation of Pharmaceutical Advertising and Promotion in Different Countries — a Critical Survey

Britain

Since 1958 the Association of the British Pharmaceutical Industry (ABPI) has attempted to regulate the promotion of prescription medicines through its voluntary Code of Practice. The Code is administered by a special committee which is chaired by an independent barrister. The committee has 14 other members — 12 drawn from the senior management of pharmaceutical companies belonging to the ABPI, and two independent physicians from outside the industry (ABPI, 1988).

The ABPI deals with matters related to the Code in three ways. Firstly, its secretariat scrutinizes advertisements in a random selection of journals, checking that they comply with the Code's provisions. Secondly, since 1985 a medical consultant to the ABPI has monitored random advertisements, examining their medical and scientific content and on occasion asking companies to submit data substantiating the claims made. Possible breaches identified in this way are followed up by informal correspondence with the company concerned. Cases not resolved informally are dealt with by the measure of last resort — referral to the Code of Practice Committee for formal investigation. Since 1983 the Committee's full adjudications have been published in *Reports to Chief Executives*, copies of which are circulated to the Department of Health and the *British Medical Journal* (Herxheimer and Collier, 1990).

The seventh revision of the ABPI Code was introduced in 1988. Its clauses deal with a detailed interpretation of the Medicines Act 1968 and its associated regulations, and with sections concerning matters of professional conduct, fair competition, 'canons of good taste' (e.g.depictions of the female form), and the reputation of the pharmaceutical industry. By international standards, the Code is quite strict, forbidding such things as: the use of 'hanging' comparatives in advertisements (e.g., claims that a product is 'better' or 'stronger' — clause 5.5); the provision of free samples of drugs except in response to a signed request from a doctor (clause 18.1); gifts irrelevant to the practice of medicine or pharmacy (e.g. table mats — clause 19.2); the extension of invitations to meetings to doctors' spouses (clause 20); and the direct advertising of prescription medicines to the general public (clause 22.2). In addition, the Code recommends that medical representatives 'should be paid on the basis of a fixed basic salary, and any addition proportional to sales of prescription medicines should not constitute an undue proportion of their remuneration' (clause 17.10). Many of these clauses follow the recommendations of the *Ethical criteria for medicinal drug promotion* published by the World Health Organization (WHO) (1988). By contrast, it is notable that few of these provisions are listed in the Code of Practice of the New Zealand Researched Medicines Industry (RMI).

Despite the detailed provisions of the ABPI Code, however, the evidence suggests that self-regulation by the industry has failed to deter promotional excesses in Britain. In a review of the complaints received by the ABPI

Code of Practice Committee between 1983 and 1988 Herxheimer and Collier concluded that the Code was frequently broken. Altogether 192 instances of breaches were found by the ABPI Committee — on average more than one breach every fortnight. In addition, an undisclosed number of complaints were made internally and handled informally, hence avoiding publication. Of the possible breaches identified by the ABPI secretariat, over 90 per cent were admitted to be breaches by the company concerned. The investigators concluded that companies were good at recognizing breaches, but that this had not led them to commit fewer of them: 'The data do not reveal any obvious deterrent effect of the code' (Herxheimer and Collier, 1990).

The self-regulation by the British pharmaceutical industry shares additional weaknesses of systems in other countries. For example, members of the general public, who are endangered by misleading or otherwise incorrect promotion, are neither informed nor consulted. Unlike countries such as Sweden and Canada (but in common with Australia and New Zealand), the ABPI Code of Practice Committee has no representation of consumer interests. Herxheimer and Collier concluded:

The ABPI's wish to secure compliance with the code seems far weaker than its wish to pre-empt outside criticism and action by showing its procedural consistency and efficiency — for example, by dealing with cases in an average time of 14 weeks. Outside complainants are given no opportunity of responding to defendants' arguments . . . [the ABPI's] self regulation seems to be a service to itself rather than to the public (Herxheimer and Collier, 1990).

The ABPI's self-regulation leaves much to be desired in other respects. For example, the ABPI gives virtually no adverse publicity to companies found to have breached the Code. The only sanction it can impose is to suspend an offending company from membership of the Association — a threat which has actually been carried out once during the last 30 years. Similarly, the ABPI has no power to enforce an offending company to publish a retraction.

In response to a report of the Royal College of Physicians (1986) entitled *Relationships between Physicians and the Pharmaceutical Industry*, the ABPI also released a position paper. Entitled *Relationships between the Medical Profession and the Pharmaceutical Industry*, the position paper made a number of recommendations aimed at increasing accountability in matters such as payment of doctors for clinical research, sponsorship of meetings, and the conduct of market research. In particular, one section entitled 'Unreasonable Requests' gives a revealing insight into the sorts of pressures exerted upon individual companies to act in a manner that could lead to breaches of the ABPI Code:

> Whilst it is recognised that pharmaceutical companies should in no circumstances make excessive payments, one of the most sensitive issues raised in any discussion on relationships between the pharmaceutical industry and the medical profession is the question of payments requested by doctors from such companies. Frank exploitation is rare, but there is a tendency for practitioners in all fields of medicine to seek increasing support from the pharmaceutical industry. Requests for funding range from reasonable payments for involvement in clinical trials through sponsorship of meetings or of attendance at meetings, both international and domestic, provision of equipment and support for non-medical events, to insistence on payment for seeing a medical representative, a strong request for provision of equipment to stock new premises and a suggestion that a company should supply its products free of charge for a stipulated period, or be boycotted by the practice concerned if it did not comply (ABPI, 1986).

Several of the recommendations made in the ABPI position paper were later incorporated into the seventh revision of its Code of Practice.

The United States

Since the passage of the Kefauver-Harris Act in 1962, prescription-drug advertising in the United States has been regulated by the Food and Drug Administration (FDA). In common with the codes of advertising practice in most other countries, pharmaceutical advertisements in the United States must be accompanied by information on precautions and listings of the product's side-effects, toxic effects, and contra-indications. This notion of 'fair balance' in drug advertisements is also embodied in the Federal Food, Drug and Cosmetic Act (1938) (Cohen, 1988; Kessler and Pines, 1990).

In countries where the regulation of pharmaceutical promotion has been left to the industry, various commentators have remarked on the restricted range of powers of sanction invested in the disciplinary committees appointed by the industry (Lexchin, 1987b; Herxheimer and Collier, 1990). By contrast, the FDA has the power to issue frequent 'regulatory letters' to companies making false or misleading promotional claims, requiring corrective advertising or a 'Dear Doctor' letter. For example, on 29 May 1991, the FDA disclosed that it had requested Bristol-Myers Squibb to stop distributing a publication that was designed to look like a scientific journal but which actually served to promote the company's cancer drugs. Bristol-Myers Squibb also agreed to send thousands of physicians a 'Dear Doctor' letter which explained the FDA's concern about the publication and 'set the record straight' on the source of the information. At the same time, the company agreed to obtain advance approval from the FDA for the next two years of all promotional materials associated with the cancer drugs (Iglehart, 1991).

Despite its broad authority over labelling and advertisements, however, the FDA has not attempted to subject all industry-sponsored activities to the strict legal requirements (Kessler, 1991). At a press briefing in February 1991, the newly-appointed commissioner of the FDA, Dr David Kessler, admitted that the agency had been a 'paper tiger' in relation to policing advertising and marketing practices in the drug industry: 'We wrote letters, we wrote letters, and we wrote more letters', he said, but the agency took few other visible actions towards enforcement (Iglehart, 1991). One official of the FDA even stated that 'the vast majority' of promotional materials submitted to the FDA are false or misleading in some respect but that the agency can take regulatory action in only about 5 per cent of cases, mainly due to the lack of resources (Anonymous, 1989). According to the Director of the United States Public Citizen Health Research Group, Dr Sidney Wolfe:

For the FDA not to have criminally prosecuted a drug company for 20 years in the face of a massive amount of violative activities is an invitation to continued law-breaking (Wolfe, 1991).

Given the lack of resources for the FDA to police its rules concerning pharmaceutical advertising and promotion, it is perhaps not surprising to note that health professionals and the pharmaceutical industry in the United States have had to develop their own codes of practice. For example, in 1988 the American Pharmaceutical Manufacturers Association (APMA) adopted its own code of ethics of marketing practices, as well as endorsing (in 1990) the American College of Physicians' guidelines regarding gifts from the pharmaceutical industry. In December 1990 the American Medical Association (AMA) adopted a set of guidelines that deem unethical the acceptance of vacations or cash, to pay for travel expenses, gifts of substantial value, cash, and lavish meals and entertainment from pharmaceutical companies — all examples of widely-practised marketing ploys in the United States, which came to light during a Senate Committee on Labour and Human Resources hearing on Promotional Practices in the Pharmaceutical Industry (Randall, 1991). Two days after the AMA released its guidelines, they were also adopted by the APMA, which has over 100 members.

There is a growing consensus in the United States that favours a substantial expansion of the authority and resources with which the FDA can carry out its increasingly complex regulatory mission (Iglehart, 1991).

The impetus for this concern appears to be the growing competition in the pharmaceutical industry which has resulted in new forms of promotional activities aimed not only at physicians but also at consumers, hospitals, and third party purchasers. These activities include holding press conferences, sponsoring media tours, encouraging media coverage of scientific symposia, issuing video news releases, sponsoring single-issue publications, and discussing drugs in development with the investment community (Kessler and Pines, 1990). In addition, direct-to-consumer advertising of prescription drugs, both in print and on television, is becoming more common. According to one commentator, the pharmaceutical industry has been inserting itself 'very subtly' into the rupture created by the breakdown in the relationship between the doctor and the patient, which meant that a better informed public was being 'encouraged to medicate' and would come to the physician demanding a cure or a pill (Mansell, 1989).

Some of the worst examples of promotional excesses came to light during the two-day Senate Committee hearings chaired by Senator Edward Kennedy in December 1990. Examples included a frequent-prescriber plan, in which Wyeth-Ayerst Laboratories gave doctors 1000 points on American Airlines' frequent-flyer programme for each patient they put on the hypertension drug Inderal LA. Among other examples cited, a consortium of 10 drug companies provided doctors with a free $35,000 computer system if they spent 20 minutes a week reviewing 'promotional messages' and 'clinical information' and completed four continuing medical-education programmes a year; Ciba-Geigy offered free Caribbean vacations to doctors in return for their sitting in on a few lectures about Estraderm, an oestrogen patch; and Roche paid doctors $1,200 if they prescribed the antibiotic Rocephin for 20 hospital patients (Purvis, 1991). Such promotional tactics probably far exceed any examples reported in other countries which mainly rely on self-regulation by the industry.

Perhaps the most disturbing element in the American system of regulation is that neither the AMA nor the APMA has any specific mechanisms in place to enforce their ethical guidelines. Gerald Mossinghoff, President of the APMA, has stated that complaints of ethics violations within the APMA have been resolved in the past in a 'policy management style'. Allegations are brought to him and then resolved with the Board within a week or two. In response, Senator Kennedy expressed concern that in the three years since the APMA adopted its Code of Ethical Conduct, not one complaint of violation had been brought to the attention of the APMA President. Regarding physician compliance with the AMA

Code of Ethics, Daniel Johnson, vice-speaker of the AMA's House of Delegates, has stated that the new guidelines will be the 'gold standard' against which physicians will be measured with respect to membership in physicians' organizations and their licence to practice (Randall, 1991).

Concluding the Senate Committee hearings, Senator Kennedy stated: 'The principal question is whether the industry and the medical profession can heal themselves or whether additional regulation or legislation is needed'. He went on to remind the hearing's audience that in 1974 the APMA President at the time — in Senate hearings on the same topic — 'expressed desire to control excesses in promotional practices. Now almost 17 years later, we have similar commitments. We're interested in seeing whether things will be different' (Randall, 1991).

Canada

The Canadian Pharmaceutical Advertising Advisory Board (PAAB) is an autonomous body formed in 1975 to provide a mechanism for the independent review and clearance of pharmaceutical advertisements prior to their release. The Board comprises equal representation from the health professions, consumers, pharmaceutical manufacturers, the professional media, the Advertising Advisory Board and non-voting observers from the pharmaceutical marketing clubs of Quebec and Ontario.

All proposed advertisements must be submitted to the PAAB for review and clearance prior to their use. PAAB-accepted advertisements are allocated an identification code comprising the PAAB logo together with a 12-month clearance. Any advertisements not bearing the PAAB logo may be notified to the PAAB for investigation. The Code also explicitly encourages health professionals, agencies, advertisers, and the media to submit advertisements to the PAAB which they consider to be in breach of the code of practice. The operating expenses of the Canadian PAAB office (including the Commissioner and secretarial staff) are met entirely through advertisement review fees, which range from $Can100-325 per submission (Godden, 1987).

The actual scope of the PAAB Code of Advertising Acceptance (PAAB, 1986) is rather limited, covering only print and audiovisual advertising media. Thus there are no provisions in the PAAB Code for the regulation of other promotional activities, such as sponsorship of meetings, hospitality, or the training of medical representatives. These are covered by the Code of Marketing Practices of the Pharmaceutical Manufacturers Association of Canada (PMAC), which is beyond the jurisdiction of the autonomous PAAB. Canada also has the Food and Drugs Act (1963) to which all pharmaceutical advertisements must conform. However, the Health Protection Branch (HPB) which is responsible for enforcing this legislation, has ceded responsibility for its enforcement to the PMAC. Officials of the HPB have opted for a co-operative and 'open door policy' with the Canadian drug companies in preference to a tough adversarial stance (Lexchin, 1990). Based on this policy, the Health Protection Branch makes only informal spot checks of drug advertising. There is no programme of continuous reviewing of advertisements to ensure that the prohibition against misleading advertising in the Food and Drugs Act is observed.

Despite the claims of success made by the PAAB about its programme of voluntary regulation (Godden, 1987), the evidence suggests that promotional excesses occur just as frequently in Canada as in other countries. In a survey of promotional practices in Canada, Lexchin concluded:

Although the industry has adopted a voluntary code of advertising practice, this has not prevented gross excess in all forms of pharmaceutical promotion: drug-company-sponsored continuing medical education, and promotion through the public media, detailers, direct mail, sampling, and journal advertising (Lexchin, 1987b).

The PAAB Code of Advertising Acceptance is so weak that advertising material which complies with it can still be deceptive. Sanctions for violations of its Code are limited to withdrawal or modification of the offending advertisement. Thus Lexchin cites one example of the powerlessness of the PAAB:

A doctor wrote to the PAAB complaining that certain irrational combination drugs were being advertised in Canadian medical journals. Although the PAAB agreed that pharmaceutically the drugs did not make sense, the PAAB stated that it could do nothing about the advertising, since the advertisements conformed to its Code (Lexchin, 1987b).

Similarly, there is no requirement for the PAAB to publicize the results of complaints against advertisements, even when these complaints have been upheld.

Australia

In 1960, the Australian Pharmaceutical Manufacturers Association (Australian PMA) became one of the first pharmaceutical organizations to endorse a voluntary code of conduct. The Australian PMA Code of Conduct (1988) encompasses many of the ethical criteria set out by the

WHO (1988). Among its positive features, the Code prohibits the promotion of prescription-only products to the general public (clause 6.2). Clause 2.9 specifies that promotional competitions must fulfil at least two of the following criteria: the competition is based on medical knowledge, or the acquisition of medical knowledge; the prize is relevant to the practice of medicine or pharmacy; individual prizes offered are of small economic value. Measured against such criteria, several competitions run by pharmaceutical companies in New Zealand would likely be found to be in breach. For example, one manufacturer of an anti-emetic ran a competition during 1990 in which local doctors were encouraged to send in suggestions for novel uses for sick bags. The prize was a case of Daniel Le Brun champagne. Another competition promoting an antidepressant drug offered a first prize of a weekend holiday for two staying at the Hyatt Hotel in Rotorua, plus a scenic flight in a helicopter 'with a luncheon stopover at beautiful Solitaire Lodge'.

However, none of the voluntary codes of practice reviewed in this chapter have taken on board all of the criteria recommended by the WHO. Thus the Australian PMA Code omits several significant provisions including:

Promotion should be in keeping with national health policies; medical representatives should not offer inducements to prescribers and dispensers; in order to avoid over-promotion, the main part of the remuneration of medical representatives should not be directly related to the volume of sales they generate; and all these bodies (government, industry, health personnel and consumers) should monitor and enforce their standards (WHO 1988).

Unlike the WHO guidelines, the Australian PMA has chosen to limit its Code jurisdiction to promotional activities that are concerned directly with drug products, thus eliminating complaints concerning alleged misleading antigeneric and antigovernment advertisements (Harvey, 1990). Although the Australian PMA Code has been subject to review, currently no provision exists for this process to include non-industry representation, as recommended by the Australian Trade Practices Commission. In spite of repeated requests, the Australian PMA has steadfastly refused to appoint representatives of consumer and/or censorial organizations, for example the Consumers' Health Forum or the Medical Lobby for Appropriate Marketing, to their review, complaint or monitoring committees. In short, according to Harvey (1990), 'the APMA [Australian PMA] has yet to accept that the price of self regulation is public accountability'.

In mitigation, the Australian PMA complaints subcommittee was reconstituted in 1988 under the chairmanship of an independent lawyer, and nominees were sought from professional associations. The committee now has representation from the Australian Society of Clinical and Experimental Pharmacologists and the Royal Australian College of General Practitioners. A representative of the Arthritis Foundation was also invited to join, presumably to address consumer concerns, 'but in practice bypassing the organised consumer movement and the accountability to members that formal consumer representation provides' (Harvey, 1990).

The Australian PMA also publicizes accounts of breaches of its Code. However, these consist of a once-only four- to five-line report of the offence in small print in the 'Noticeboard' of the *Medical Journal of Australia*, which is published six to nine months after the breach has occurred. The other sanction applied by the Australian PMA to offending companies is to request that the advertisement be withdrawn or modified before further use. In practice, given that the complaints procedure takes a number of months to produce a result, these advertisements have usually run their course (and have achieved their effect) by the time the sanction is applied. Consistent with surveys of self-regulation in other countries, a recent reviewer concluded:

There is no evidence that the above 'sanctions' have deterred companies from repeated Code breaches. For example, Smith Kline & French Laboratories have been associated with nine Code breaches over a 12-month period, in spite of 'admonishment' by the APMA [Australian PMA] and condemnation by health professionals and consumer groups. Smith Kline & French Laboratories also has highlighted the fundamental flaw in the APMA's self-regulatory system — faced with sanctions it finds unacceptable, a company can simply resign from the APMA and evade the penalties (Harvey, 1990).

Sweden

In Sweden almost all pharmaceutical companies belong to either the Association of the Swedish Pharmaceutical Industry or the Association of the Representatives of Foreign Pharmaceutical Industries. In 1969 these two organizations established *Rules for Information about Prescription Drugs*, and at the same time set up a commission to monitor the firms' adherence to the guidelines and to rule on complaints. The commission has a chairman, a vice-chairman and a secretary, all of whom must be lawyers with expertise in marketing law. There are in addition 10 members consisting of six representatives from drug companies, two medical doctors, and two members representing the public interest. The commission meets about eight times each year, and its deliberations are decided by majority vote (Perman, 1990).

The commission's activities include a 'watchdog' function, which ensures that printed information (mainly advertisements and mailings) to health professionals is monitored regularly. The commission's activities are entirely financed by the Swedish pharmaceutical industry. Where a company is found to have breached the industry's code of practice, the commission requests the offending promotional activity to be altered or terminated. Accounts of breaches are published in the *Journal of the Swedish Medical Association* as well as in *Scrip*, the news-magazine of the international pharmaceutical industry. Finally, the offending company is also required to pay a handling fee (about $US5,000) to the commission.

Does the Swedish system work? The fact that no decision made by the commission has yet been appealed has been cited as evidence to suggest that the system has gained the respect of those who use it. That the Swedish National Board of Health has used the services of the commission has been taken as further evidence that government authorities consider it both competent and unbiased. Nonetheless, certain promotional practices have been approved by the Swedish commission that might not be condoned by regulatory authorities in other parts of the world. For example, the commission was asked for an opinion on whether it was acceptable to directly promote drugs by mail to persons under the age of 18 years. One company used such mailings in the promotion of a drug for dysmenorrhoea. In this instance, the commission chose not to specify a 'general lowest age limit', but emphasized that such drug promotion to young members of the public required 'particular care and caution'. There are additional aspects of pharmaceutical promotion not covered by the commission's terms of reference, for example there are no guidelines at present regarding the remuneration of doctors for their participation in the marketing activities of pharmaceutical companies (Perman, 1990).

New Zealand

The role of the New Zealand Government as a residual regulator of pharmaceutical promotion is no exception to the pattern seen in other countries surveyed so far. Despite the wide powers of enforcement vested in the Minister of Health under the provisions of the Medicines Act (1981) and the Medicines Regulations (1984), the Health Department by its own admission lacks the necessary resources or personnel to monitor pharmaceutical advertising and promotion (Boyd, 1990). Accordingly, the policing of promotional standards is left to the industry and to the health professions.

Like its counterparts in other countries, the Researched Medicines Industry Association (RMI (formerly the Pharmaceutical Manufacturers

Association (PMA)) has, since 1962, administered its own code of advertising practice. The eighth revision was adopted in September 1990, with the stated objective of ensuring the provision of 'accurate, fair and objective information on medical products' to the medical and allied professions 'so that rational prescribing decisions can be made' (RMI, 1990). The RMI Code covers a wide range of topics including the use of printed promotional material, direct mailings, audio-visual material, drug samples, medical representatives, gifts and inducements, hospitality, and market research. In common with counterparts in other parts of world, the RMI Code also suffers from imprecise use of language; for example '[drug] sample distribution is . . . to be reasonable in quantity'; hospitality 'should be kept within reasonable bounds and would normally be directed towards members of the medical and allied professions'; and in conducting market research 'any incentives offered to the interviewees should be kept to a minimum and be commensurate with the work involved'.

Whether the latest revision of the RMI Code lives up to its promise of ensuring the delivery of 'accurate, fair and objective' information remains to be seen. Previous analyses of printed advertisements in New Zealand have found that earlier versions of the Code were frequently breached (Lexchin, 1987a; Peacock, 1990). These breaches occurred despite the assurances of the former Executive Director of the PMA that:

PMA member companies are aware that every time a company falls prey to the temptation to use questionable marketing techniques as a competitive weapon, it risks a serious blow to its reputation and prestige. This, in itself, is a strong restraining influence on marketing behaviour (Hardy, 1987).

It is a credit to the RMI that the latest version has expanded and improved the sections regarding the enforcement of the Code by its Disciplinary Committee. The range of sanctions applicable to infringing companies are now clearly spelt out, and the disciplinary procedures appear to have become much more explicit. Previously, the Disciplinary Committee consisted of only three members comprising the Chairman (who was a legal adviser to the RMI or a Barrister of the High Court), the President of the RMI or his nominee, and one other member of the RMI. In the latest revision, membership of the Disciplinary Committee has been somewhat expanded to include (in addition to the Chairman) one practising member of the medical profession who is not an employee of an RMI-member company, two members of the RMI Board who are not involved with the complaint, and a Medical Director from a member company.

As in the previous versions of the Code, the Disciplinary Committee meets with 'such frequency as the amount of business requires'. Thus the enforcement of the Code still relies to a large degree on individual companies conforming, with the RMI only becoming involved once a complaint has been made. The range of sanctions which the RMI Board may exercise include: requirement for the infringing company to discontinue an offending promotional practice; issuing of retraction statements by the member; suspension or expulsion of the member from the RMI; ordering the member to pay the costs and expenses of the inquiry; and publication of the salient points of the breach in the six-monthly report of the Board. Broader dissemination of the breach will be at the discretion of the Board, and is by no means automatic. Indeed, according to the previous Chief Executive of the RMI, Bill McLauchlan, such information is 'probably not of great concern to the consumer' (Consumers Institute, 1990). The actual extent to which the RMI Board is willing to apply the range of sanctions spelt out in the new version of the Code is a matter for speculation — as is the degree to which the same sanctions will discourage member companies from breaching the Code.

Regarding the direct advertising of prescription medicines to the general public, this matter is left to the discretion of each individual company. The Code recommends that companies give 'careful consideration . . . to all the ramifications before reaching a decision to do so' (RMI, 1990). During 1989, the direct advertising of prescription medicines to the New Zealand public by various drug companies attracted critical comment on several occasions. In one instance, Smith Kline and French placed a two-page colour advertisement in the *Listener* magazine promoting vaccination against hepatitis B. The drug company's name was not included in the advertisement, which provoked criticisms that the public might take the advertisement to be part of the Health Department's vaccination programme. On a further occasion, immediately following the publication of a case control study which linked asthma mortality to the bronchodilator fenoterol, Edinburgh Pharmaceutical Industries bought newspaper space to make a public statement about the reliability and efficacy of their rival product Ventolin (Coney, 1989).

Neither the Medicines Act (1981) nor the Medicines Regulations (1984) prohibit direct-to-the-public advertising of prescription medicines. However, the legislation does require advertisements for medicines to include information about the drug's contra-indications, adverse effects, and a clear statement that the medicine is 'restricted' or 'prescription only'. In both the examples cited above, the regulations appeared to have been breached, prompting the Health Department to take its concerns to the manufacturers. Smith Kline and French argued that its advertisement did not promote its product, but nonetheless agreed to withdraw the advertisement (Coney, 1989). Murray Main, the commercial manager of Glaxo (of which Edinburgh Pharmaceuticals is a group company), has gone on record to state that their advertisement was aimed at reassuring Ventolin users that their medication was not implicated in the fenoterol scare; nonetheless, he admitted that the advertisement 'did not entirely meet regulations' (*Manawatu Evening Standard*, 29 December 1990).

The recent New Zealand experience with direct-to-the-public advertising of prescription medicines illustrates just one of the multitude of problems confronting the regulation of advertising and promotion in an increasingly competitive industry. Some have argued that direct-to-the-public advertising meets the increasing consumer demand for health information, alerts consumers to new treatments, encourages people to seek medical advice for conditions that would otherwise go untreated, and generally results in a more informed public (Masson and Rubin, 1985). On the other hand, it has been argued that such advertising interferes with the doctor-patient relationship, confuses patients, increases the cost of drugs, puts undue emphasis on pharmacological treatment alternatives, pressures physicians to prescribe drugs, and results in unnecessary use (Cohen, 1988).

In the United States, the FDA has repeatedly declared that direct-to-the-public prescription advertising is not in the public interest (Cohen, 1988). In 1983, the FDA requested a voluntary moratorium on direct-to-consumer advertising so that it could explore the issue. Nonetheless, in 1985, the agency lifted its moratorium, declaring that its current rules were sufficient to regulate the area (Kessler and Pines, 1990). Presently, the FDA's rules do not specifically forbid such advertisements (Cohen, 1988; Kessler and Pines, 1990). Continuing to permit this form of advertising has led to further increases in the ratio of advertising to sales, already high in the drug industry. Direct advertising thus increases costs (to the consumer or the taxpayer) without necessarily promoting competition (Cohen, 1988).

In addition to self-regulation by the industry, the medical and allied health professions in New Zealand have developed their own guidelines *vis-à-vis* their relationship with the pharmaceutical industry. Guidelines have been issued by the New Zealand Council for Postgraduate Education (1983); the Wellington Faculty of the Royal New Zealand College of General Practitioners (Durham, 1990); the New Zealand Hospital Pharmacists' Association (1990); and the New Zealand Medical Association

(NZMA) (1990). However, lacking specific mechanisms for monitoring, enforcement and sanctions, most of these guidelines have only the status of position statements. Furthermore, many of the professional guidelines share the same weaknesses of their overseas counterparts, particularly in their use of vague and discretionary language. For example, the position statements issued by the NZMA in September 1990 discourage the acceptance of 'excessive' gifts, amenities and hospitality from the industry, and urges doctors to participate in practice-based trials of pharmaceuticals conducted in accordance with 'basic precepts of accepted scientific methodology' (NZMA, 1990). In the absence of stricter definitions of 'acceptable' practice, the interpretation of such guidelines is most often left to the individual doctor. Thus, according to the NZMA Manawatu branch secretary, Dr Donald Fulton: 'Most doctors would think a rafting trip during a study weekend was okay, but they'd be less likely to accept a rafting trip on its own. It's basically a commonsense issue' (*Manawatu Evening Standard*, December 1990).

Conclusion

The failure of self-regulation to curb promotional excesses is an indictment of Health Ministers in all the countries surveyed. By failing to enforce the regulations controlling drug promotion, they have successively abrogated their responsibilities to the pharmaceutical industry and the health professions. There is no country yet in which promotion is satisfactorily controlled by self-regulation (Herxheimer and Collier, 1990).

There is an inherent conflict between the promotional advertisement of drugs to increase sales, and the provision of adequate and objective drug information to facilitate rational prescribing (Cohen, 1988). In a survey of drug advertising, Krupka and Vener (1985) reported that the greatest investment in advertising appeared to be required to achieve high levels of sales for drugs which did not have a clear-cut ameliorative effect. That drugs of lesser benefit require more advertising than those whose benefit is known, confirms the premise that advertising seeks to increase sales, not knowledge (Cohen, 1988).

Steps to Improve Standards in Pharmaceutical Promotion

Codes of advertising practice

No single code of practice reviewed in the present chapter satisfied all of the WHO-recommended ethical criteria for pharmaceutical advertising. To be effective, codes of practice must be comprehensive in their coverage of promotional practices, and in addition must be subject to review and amendment on a regular (e.g. two-yearly) basis in order to respond to the evolving nature of advertising tactics. Loopholes are evident in virtually all of the codes currently operating in different countries, and experience has shown that pharmaceutical companies will exploit these weaknesses for competitive advantage.

Perhaps the most important requirement of pharmaceutical promotion ought to be that it is in keeping with each country's national health policies, as recommended by the WHO (1988). Such an overriding principle may then lead to the restriction of promotional tactics (such as direct-to-the-public advertising of prescription drugs, or the free sampling of drugs which are as yet unlisted on the Drug Tariff) which tend to undermine legitimate Government efforts to rationalize other aspects of pharmaceutical utilization, such as approving products for marketing, monitoring adverse reactions, or regulating prices.

Monitoring infringements

Having established an enforceable code of practice, clearly it is unsatisfactory to abandon its policing to a haphazard process of random checks made by harassed government officials or to sporadic complaints raised by rival companies. Experience shows that a system left to police itself will lead to a proliferation in promotional excesses. In a system which provides for a disciplinary body to meet only *after* a complaint has been lodged, there will always be a disincentive to 'play by the rules'; it seems far easier for a company to match its rival by repeating the original breach. Given this experience, the Canadian PAAB provides a more satisfactory model to ensure that breaches do not arise in the first place. The strength of the PAAB lies in its ability to independently vet all promotional material before it is used. Prospective clearance of advertisements would ensure that all companies compete on a 'level playing field'. In order to be workable, the Canadian model would need to be extended to other areas of promotion, not just restricted to printed media.

Sanctions

The inadequacy of sanctions partly explains why self-regulation has failed to deter promotional excesses in the countries studied. In the case studies included in this chapter, formal mechanisms were often not identified for publicizing the nature of the breaches. The former Executive Director of the New Zealand PMA was probably accurate in stating that a strong restraining influence on the behaviour of companies is the risk of a serious blow to their reputation and prestige resulting from a questionable marketing tactic. Unfortunately, in New Zealand, as in several other countries, there are no channels by which the consumers of a company's products (namely the doctor and the patient) can be automatically notified of complaints that are upheld. What the consumers don't know can't hurt the company. Even in countries such as Australia which do publicize details of infringements, the notices are inconspicuously posted long after the breach has occurred, thus diminishing the impact of the association. The industry has hitherto not lacked in innovative methods to circumvent its own codes of practice; it would not be inappropriate then to request industry to devise its own effective sanctions.

Public accountability

Finally, to dispel criticisms that voluntary regulation is a self-serving exercise, the pharmaceutical industry ought to be encouraged to include consumer representatives in its disciplinary process. The industries in some countries reviewed in the present chapter already have consumer representation. As the ultimate consumer of the industry's products, the general public has a legitimate interest in the ways in which companies deal with the medical profession. Public accountability could only enhance the industry's image as a responsible member of society. Certainly there seems to be little justification for the attitude encapsulated in the following advice to the industry from a public relations firm: 'Events in recent months in both Australia and New Zealand have highlighted the potential threat which consumerism poses to the pharmaceutical industry' (Network Communications, 1986).

In conclusion, despite incremental improvements in voluntary regulation by the pharmaceutical industry, deficiencies still remain in virtually all of the countries reviewed. The failure of self-regulation to curb promotional excesses has prompted public and professional concern in several countries (Kawachi, 1990; Harvey, 1990; Herxheimer and Collier, 1990; Randall, 1991). In an era of rising drug utilization and expenditure, the question of how long governments can afford to remain regulators of the last resort remains to be determined.

References

Anonymous (14 February 1989) FDA's drug promotion problems *Scrip* 24.

Association of the British Pharmaceutical Industry (ABPI) (1986) *Relationship between the medical profession and the pharmaceutical industry* London, Association of the British Pharmaceutical Industry.

Association of the British Pharmaceutical Industry (ABPI) (1988) *Code of Practice for the Pharmaceutical Industry*, seventh edition. London, Association of the British Pharmaceutical Industry.

Australian Pharmaceutical Manufacturers Association Inc. (APMA) (1988) *Code of Conduct of the Australian Pharmaceutical Manufacturers Association Inc.* Sydney, Australian Pharmaceutical Manufacturers Association Inc.

Boyd, B. (1990) What you see depends on where you sit — Department of Health perspective. In Kawachi, I. (ed) *Pharmaceutical Advertising and Promotion — Options for Action* 50–51, Wellington, Public Health Association of New Zealand.

Cohen, E.P. (1988) Direct-to-the-public advertisement of prescription drugs *New England Journal of Medicine* 318: 373–5.

Coney, S. (1989) Pharmaceutical advertising *Lancet* i: 1128–9.

Consumers Institute (April 1990) Undue influence. How drug companies sway doctors' decisions *Consumer Food & Health*: 7–9.

Durham, J. (1990) Pharmaceutical promotion in New Zealand — A view from general practice. In Kawachi, I. (ed) *Pharmaceutical Advertising and Promotion — Options for Action* 43–49 Wellington, Public Health Association of New Zealand.

Godden, J.O. (1987) Monitoring drug advertisements *Lancet* i: 980.

Hardy, A.J. (1987) Drug promotion *New Zealand Medical Journal 100: 667.*

Harvey, K. (1990) Pharmaceutical promotion *Medical Journal of Australia* 152: 57–8.

Herxheimer, A. and Collier, J. (1990) Promotion by the British pharmaceutical industry, 1983–1988: a critical analysis of self regulation *British Medical Journal* 300: 307–11.

Iglehart, J.K. (1991) The Food and Drug Administration and its problems *New England Journal of Medicine* 325: 217–20.

Kawachi I. (ed) (1990) *Pharmaceutical Advertising and Promotion — Options for Action* Wellington, Public Health Association of New Zealand.

Kessler, D.A. and Pines W.L. (1990) The federal regulation of prescription drug advertising and promotion *Journal of the American Medical Association* 264: 2409–2415.

Kessler, D.A. (1991) Drug promotion and scientific exchange. The role of the clinical investigator *New England Journal of Medicine* 325: 201–203.

Krupka, L.R. and Vener, A.M. (1985) Prescription drug advertising: trends and implications *Social Science and Medicine* 20: 191–7.

Lexchin, J. (1987a) Advertisement scrutiny *Lancet* i: 1323–4.

Lexchin, J. (1987b) Pharmaceutical promotion in Canada: convince them or confuse them *International Journal of Health Services* 17: 77–89.

Lexchin, J. (1990) Drug makers and drug regulators: too close for comfort. A study of the Canadian situation *Social Science and Medicine* 31: 1257–63.

Mansell, P. (10 November 1989) Drug advertising — difficult decisions ahead *Scrip* 1463: 21–3.

Masson, A. and Rubin, P.H. (1985) Matching prescription drugs and consumers *New England Journal of Medicine* 313: 513–5.

Network Communications (November 1986) Consumerism challenges the industry *Communicator* 5.

New Zealand Council for Postgraduate Medical Education (1983) The relationship between postgraduate organizations and institutions in New Zealand and pharmaceutical companies *New Zealand Medical Journal* 96: 578–9.

New Zealand Hospital Pharmacists' Association (April 1990) Guidelines for the conduct of pharmaceutical industry representatives in hospitals.

New Zealand Medical Association (NZMA) (1990) *Position statements on the relationship between the medical profession and the pharmaceutical industry* Wellington, New Zealand Medical Association.

Peacock, S. (1990) Two surveys of pharmaceutical promotion and advertising. In Kawachi, I. (ed) *Pharmaceutical Advertising and Promotion — Options for Action* 54-6, Wellington, Public Health Association of New Zealand.

Perman, E. (1990) Voluntary control of drug promotion in Sweden *New England Journal of Medicine* 323: 616-7.

Pharmaceutical Advertising Advisory Board (PAAB) (1986) *PAAB Code of Advertising Acceptance* Ontario, Pharmaceutical Advertising Advisory Board.

Purvis, A. (18 March 1991) Cheaper can be better *Time Magazine* 60.

Randall, T. (1991) Kennedy hearings say no more free lunch — or much else — from drug firms *Journal of the American Medical Association* 265: 440-2.

Researched Medicines Industry Association of New Zealand (RMI) (November 1990) *Code of Practice*, eighth revision, Wellington, Researched Medicines Industry Association of New Zealand Inc.

Report of the Working Party of the Royal College of Physicians (London) (1986). The relationship between physicians and the pharmaceutical industry *Journal of the Royal College of Physicians of London* 20: 235-57.

Wolfe, S.M. (1991) Promotional practices in the pharmaceutical industry *Health Action International News* 59, 1-2, 10-11.

World Health Organization (WHO) (1988) *Ethical criteria for medicinal drug promotion* Geneva, World Health Organization.

A Guide to Practice

Peter Davis

Introduction

The chapters in this book have set pharmaceuticals and the pharmaceutical industry in a wider policy context. In this respect New Zealand is just a microcosm of forces that are at play worldwide, although perhaps the developments of the past decade have presented these issues in a particularly clear and distilled form. What, then, are the essential forces shaping the policy issues addressed in this book?

Firstly, the international pharmaceutical industry is facing considerable financial uncertainty in the longer term. While the industry still records high profits and cash flows, changes in its environment since the early 1960s have significantly reduced returns on new investment and have increased the level of risk. A combination of tighter safety regulations, diminished probabilities of commercial success, and depleted research opportunities has seen 'a marked shift in the risk/return profile of investment in the pharmaceutical industry' (Hill and Hansen, 1991). This new market reality gives the activities in New Zealand of the research-based companies and of their association a particular urgency, even a certain stridency.

Secondly, governments everywhere are confronting fiscal problems. This is an issue of more general application — the so-called 'fiscal crisis of the State' (O'Connor, 1973) — but it has found a particular expression in the English-speaking countries of the world where governments have moved to reduce State intervention and increase the scope of market forces (Castles, 1990). This combination of fiscal pressure and deregulatory zeal is exemplified in the pharmaceuticals area in New Zealand.

Thirdly, society is having to grapple with an ever more complex set of issues concerning the control and use of technology. The rapid pace of development and the pervasive influence of new technologies are taxing conventional means of arbitrating the many ethical, financial, social, and political issues raised by scientific innovation. Pharmaceuticals are an illustration of both the strengths and weaknesses of our response to new and powerful technologies.

Finally, the traditional role of the doctor as an autonomous professional exercising independent judgement is increasingly under threat, not from an interventionist and bureaucratic State but from the burgeoning competitive pressures of the health care industry (Relman, 1991). Recent events in New Zealand have underlined the failure of traditional mechanisms of peer review and collegial scrutiny (Committee of Inquiry, 1988). The pharmaceutical industry epitomizes the competitive pressures of the broader health care field on the practice of medicine.

What's to be Done?

The set of policy problems generated by these four underlying pressures — on the industry, the State, the public, and the profession — have been canvassed extensively in this book. They suggest a range of institutional changes, a number of which have been widely discussed in New Zealand, some of which are under way, and others that have not been given the serious consideration they deserve.

The State

On the part of the State, a major policy objective must be to continue to both simulate and stimulate price competition where the structure of the pharmaceuticals market permits it. Such conditions exist where drugs can be classified into clearly defined therapeutic groups — that is, groups of drugs with the same broad therapeutic indications for use and outcome (Coopers and Lybrand, 1986). By imposing part-charges and by reimbursing at the price of the lowest-cost equivalent in a therapeutic group, the State is forcing an element of price competition onto the market that otherwise would not be there. This is currently subject to litigation. In the case of non-therapeutic group drugs, licensing and direct importation are bargaining counters. Again, the issue is before the courts.

Secondly, it has been suggested that New Zealand should embark on international co-operation in registering new medicines (Edwards, 1989). This trend should be greatly speeded up and applied more broadly, not only to new drugs but to modifications for existing drugs, to generics, to monitoring for adverse reactions, and so on. As the Business Regulation Review Unit (BRRU) stated in its review of therapeutic goods regulations in Australia: 'There are . . . many savings to be made in the evaluation of new pharmaceutical entities if greater reliance is placed on overseas assessments. It is hardly credible for Australia to claim that its scientists and technologists are superior to those of comparable countries' (BRRU, 1989: 21). A subsequent report by Dr Peter Baume pointed out that Australia could not hope to match the rigour of assessment of key overseas regulatory agencies and recommended that the routine evaluation of individual patient data cease 'forthwith' (Baume, 1991: 36Z-7). The report also recommended that declarations of any conflicts of interest by members of the evaluation committee should be on the public record and available to interested parties (Baume, 1991: 80).

A move to greater reliance on international co-operation would free up resources for what are, arguably, the more cost-effective tasks of price negotiation, evaluation of drugs for listing on the Drug Tariff, and scrutiny of performance after market release. Approval delays would be reduced, New Zealand would not run the risk of pioneering drugs ahead of rigorous international scrutiny, and there could be a greater interest in the harmonization of information on products for the purposes, variously, of price negotiations, of competitive importation, and of post-marketing surveillance.

The Professions

In the case of the professions, both pharmacy and medicine could consider important institutional changes. In order to enhance the professional independence of pharmacy, for example, it would be important to separate the remuneration of the pharmacist from the cost of the drug being dispensed. Hence, as suggested in the Coopers and Lybrand report, the pharmacist should be refunded only for the actual cost of the drug, plus a dispensing fee. This would remove the link between the pharmacist's remuneration and the cost of the drug, and would also ensure that the benefits of competition achieved through industry discounting flow to the patient and the State rather than to the dispenser. Competition could also be permitted in professional fees between pharmacies, and restrictions on ownership could be removed. An extension of the range of drugs available direct from the pharmacist without prescription would also enhance the pharmacist's professional role, as well as reducing costs to the patient (Oster et al., 1990; Ryan and Yule, 1990).

The quality of the professional ethic is also a matter at issue in the case of medical practice. The Coopers and Lybrand report argued that there is wide variation in the nature and quality of prescribing, that at least part of this is likely to be due to inappropriate prescribing, and that levels of prescribing are related to the availability of doctors rather than to need. This phenomenon of medical variation is addressed in both a local

(Malcolm and Higgins, 1981; Davis, Lay Yee, and Millar, 1991) and an international literature (Andersen and Mooney, 1990) and raises serious questions about the consistency and scientific and professional basis of clinical decision-making. There are two major problems here: the circumstances that make it difficult for *individual* doctors to make the necessary comparative judgements in selecting drugs, and the lack of an effective mechanism for achieving a *professional* consensus on the appropriate selection of medicines.

The difficulties that face individual doctors in trying to make comparative judgements are many. Firstly, of course, there is the simple matter that doctors are often making complex clinical judgements at speed and under conditions of considerable uncertainty. Secondly, they are inundated with subtle and persuasive messages and inducements from the pharmaceutical companies. Thirdly, there is little comparative data actually available for ready reference (Millar, 1989). Finally, there is the sheer mathematical impossibility of gaining sufficient experience with a given drug to assess its performance. Temin (1980: 114–5) has estimated that the average doctor writes a *new* script for a given drug only about a dozen times a year, depending on patient throughput and on the range of drug products used. This average is fairly constant because the workload and the product range are correlated, with low activity doctors tending to use fewer drugs and high activity doctors using a wider range. Even the fact that for any given doctor some drugs are used more than others merely underlines the point; for any given workload, greater experience with a subset of drugs has to be bought at the expense of *less* experience in the use of the remainder. As Temin concludes: 'Doctors who use a drug a dozen times in the course of a year cannot evaluate its benefits and costs relative to those of other drugs even if they try'.

It is in recognition of this simple mathematical fact, if for no other reason, that leaders in the profession in several countries have taken seriously the idea of establishing a prescribing formulary (set of guidelines) for practitioners. The Swedes, for example, have developed such a formulary in conjunction with a programme of continuing medical education. In future, a competent general practitioner in Sweden will be expected to have expertise in clinical pharmacology at three levels of performance: detailed knowledge of about 50 drugs used for the commonest conditions embracing much of the routine workload of general practice; competent familiarity with drugs used for less common conditions (about 150 drugs); and limited knowledge of all remaining drugs such that the relevant authoritative reference can be studied on use (McGavock, 1990). It is also worth noting that the process of compiling such a formulary in a country like New Zealand could assist in the restructuring of the Drug Tariff.

Unfortunately, the simplicity and persuasive logic of this concept founder on the second major problem identified above — the lack of a mechanism for achieving a professional consensus on matters of clinical decision-making. The ethic of cinical freedom is strong, and this, together with the organizational autonomy of medical practice, defeats attempts to formulate and establish guidelines.

This failure is particularly crucial in the pharmaceutical area because in the absence of any formal professional consensus it is the industry that steps readily into the void; it sets the agenda, supplies the information, and resources the essential educational and research activities. This might merely be matter for regret, were it not for the more far-reaching consequences of such a development; the danger is that the lavish and unremitting wooing of the *individual* practitioner erodes the ethical and scientific foundations of the *profession*. The guidelines established for relationships with the pharmaceutical industry are so consistently and openly violated that the integrity of the wider ethical framework on which the profession is based is seriously threatened.

The Public

This book has canvassed a number of options in the reform of the institutional arrangements governing the relationship between the pharmaceutical industry and the medical profession and, more generally, the wider society. The issue of marketing and promotion epitomizes this relationship. This book has argued that the commercial motivation of the industry needs to be taken seriously, that arrangements should be instituted that bring the wider public interest to bear in a powerful and decisive fashion, that proceedings should be conducted openly with evidence and counter-evidence traded in public, and that consumers be involved centrally at every point.

Some solutions to the problems of commercial intrusiveness are breathtakingly simple. For example, the American Hospitals Association (AHA) has established a 'Pharmaceutical Roundtable' to which participating companies make substantial contributions. This then provides a fund, at arm's length, for genuinely independent and peer-reviewed research. In return, the companies are acknowledged — either individually or as a group — as subscription members in AHA publications and at scientific meetings, and their contributions are acknowledged in any papers that arise from such research (Anon., 1990). Similar arrangements could be developed more widely, once it had been established that the intentions of the industry were seriously to inform the profession and the public rather than merely to persuade.

The question of advertising touches upon a more subtle issue of much wider significance that has received considerable attention in this book, albeit more usually by inference than by direct allusion. Just as consumers are largely absent from deliberations on the production, distribution, and marketing of pharmaceuticals, so the more subtle aspects of the social impact of pharmaceuticals are rarely considered in the important forums of the profession and the industry.

Are we inevitably and inexorably to be a 'medicated' society? Is the best working model of the person a biologically reductionist prototype that interprets the various frailties of the human condition in terms of receptors, biochemical pathways and hormone responses? Should the biomedical pioneers of the clinic, the laboratory bench, and the corporate boardroom be permitted to promote, without ethical restraint and wider debate, a biomechanical model of the human species that has such potentially far-reaching social consequences as, for example, the routine and widespread application of hormone replacement therapy?

Conclusion

What the narrower issue of pharmaceutical promotion, and the broader question of social impact, highlight are the deficiencies of the conventional medical model — at the level of both the individual practitioner and the profession. The industry is something more than a disinterested partner for the individual clinician struggling to make difficult therapeutic decisions; it has powerful commercial motivations. By the same token, pharmaceutical therapies are not devoid of moral and social colouring; they cannot be treated by the profession merely as routine and neutral interventions in the minds and bodies of an undifferentiated human species. In these circumstances the traditional model of medical practice is not a helpful guide.

There is an opportunity here to develop a model of medicine that is more sophisticated, that is informed by the discourse of the policy sciences, that is philosophically reflective, and that, ultimately, takes more seriously the essentially humanist impulse of medical practice.

Index

Abbott Laboratories 259
Adis Press Ltd 250
ageing
  women's attitude to 182–6, 195
  skin 185
Allen and Hanbury 235, 257
Amarant Trust 186
American College of Physicians
  guidelines re gifts 273
American Medical Association
  Code of Ethics of 273, 275
American Pharmaceutical Manufacturers
    Association
  political pressure by 65–6
  Code of Ethics of 273–5
Association of the British Pharmaceutical
    Industry
  Code of Practice 269–72
Association of the Representative of
    Foreign Pharmaceutical Industries
    (Sweden) 278–9
Association of the Swedish Pharmaceutical
    Industry 278–9
asthma
  adverse reactions research 75–8
  see also fenoterol
Asthma Task Force 78–9, 81–2
Astra 264
Australian Menopause Society 199
Australian Pharmaceutical Manufacturers
    Association
  Code of Conduct of 276–8
Ayerst Laboratories 185, 188, 192–5,
    198–201

bendrofluazide 151
beta-blockers 147
bioequivalence 63
blood cholesterol, high *see*
    hypercholesterolaemia
blood pressure, high *see* hypertension
Boehringer Ingelheim
  and fenoterol 70–2, 82–93
brand name drugs 20, 25, 29
  and commodity generics 233
  substitution 235–42
  New Ethicals Catalogue 251
breast cancer
  Depo-Provera 130–6

oestrogen 193–200
Bristol-Myers Squibb 272
British Medical Association, New Zealand
    Branch 19
British Pharmaceutical Codex 20
British Pharmacopoeia 20

Canadian Pharmaceutical Advertising
    Agency Board 275–6
cancer *see*
  endometrial cancer
  cervical cancer
  breast cancer
  gallbladder cancer
Capoten 172, 252
captopril 172, 173, 253, 257
cardiovascular disease
  and oestrogen 193, 197–200
cervical cancer
  Depo-Provera and 132–3
Chemists' Service Guild 19, 28
cholesterol, high *see* hypercholesterolaemia
cholestyramine 154
Ciba-Geigy 186, 192, 200–2, 233, 274
  transdermal oestrogen patches 186
Clinical Services Letter 28, 248–9
clofibrate
  and WHO trial 144, 152
consumer empowerment 57–8
contraceptives
  marketing 98–9
  oral 100, 112–13
  transdermal implant 138
  injectable 119–41
  see also Depo-Provera; intra-uterine
    devices; Norplant
Control of Prices Act 23
Copper 7 *see* intra-uterine devices
Coronary Drug Project 153
coronary heart disease *see* cardiovascular
    disease; hypertension;
    hypercholesterolaemia

Department of Health
  pharmaceutical benefits 40–4, 48–9
  prescription costs 42, 44, 53–60, 67–70,
    202, 214, 233
  generic drugs 48, 60–73, 237–41

Pharmaceuticals Review
  Strategy 55–72
  fenoterol 70–4, 81–92, 95–6
  Copper 7 103, 105, 107–12
  regulation of drugs 114, 135–7
  Depo-Provera 122, 131, 135
  drugs assessment 216–17, 232
  information services 246–9
Depo-Provera 119–40
  clinical trials 120
  adverse reactions 124
  bone density 127–30
  breast cancer 130–1, 140
  WHO breast cancer study 131–2, 140
detailers *see* pharmaceutical industry,
  representatives
diuretics 147, 174
Douglas Pharmaceuticals 241, 257
DMPA *see* Depo-Provera
drug evaluation 59
drug information 245–64
  Department of Health 246–9
  commercial sources 249–61
drug labelling 5
drug patents *see* patents, drug
drug safety control *see* pharmaceuticals,
  regulation
Drug Tariff 20, 29, 59, 110, 111, 126,
  137, 221, 251, 292
drug treatment protocols 57
drugs *see* brand name drugs; generic
  drugs; drugs by specific brand or
  generic name; pharmaceuticals;
  prescriptions
drugs, adverse effects 95–6
  see also Medicines Adverse Reactions
  Committee; specific drugs by brand
  or generic name

Edinburgh Pharmaceutical Industries
  281–2
Eli Lilly 233, 259
enalapril 173
endometrial cancer
  and oestrogen 183, 193–4
Evans Medical Ltd 233

Family Planning Association
  Depo-Provera and 112, 124–8, 132,
  140
FDA *see* Food and Drug Administration,
  US
fenoterol 70–3, 75–96
Food and Drug Administration, US 77,
  101–4, 119–21, 124–6, 127, 130, 138,
  199, 239–40, 253, 272–5, 282
  Copper 7 101–4, 109
  Depo-Provera 119–21, 124–6
  Premarin 199

generic drugs 239–40
Friendly Societies' Medical Institute 19

gallbladder cancer
  and oestrogen 193
G.D. Searle Ltd 100–13, 188, 233
  Copper 7 100–13
Genentech 264
general practitioners
  fees of 19, 263–4
  prescribing practices of 211–27, 245–6,
  263, 290
  and treatment options 215–16
  and pharmaceutical trials 217
  and health expenditure 220
  legal and ethical considerations 221–7
  influence of advertising on 223–7
  and pharmaceutical representatives 255
  see also medical practitioners
generic drugs 10, 48–50, 58, 60–70, 229,
  232–42
Glaxo Laboratories (NZ) Ltd 22, 233,
  235, 282
Gravigard 103, 108–12

Halcion 136
Hardie Marks and Associates 69
Health Consulting Group 70
Helsinki Heart Study 153
hormone replacement therapy 179–203
  promotion 183–6, 201–3
  transdermal oestrogen patches 186, 192
  costs 202
  see also cardiovascular disease;
  osteoporosis; Premarin
hydrochlorothiazide 174
hypercholesterolaemia 144
  risk factors 152–8
  treatment guidelines 155–6
  mass treatment 156–8
hypertension 144, 146–52, 162–75
  risk factors 146–52
  treatment guidelines 154–6
  mass treatment 156–8, 168–72
  drug marketing 164–6, 172–5
  mortality 166–7
Hypertension Detection and Follow-up
  Program 150

International Federation of
  Pharmaceutical Manufacturers
  Associations 5
International Health Foundation 189
Intra-uterine devices 98–114
  Lippes loop 99
  Dalkon shield 99, 100, 104
  Copper T 99
  Copper 7 100–12
  adverse effects 102–4

clinical trials 102-3
isoprenaline 76-7
IUD *see* intra-uterine devices

Johnson and Johnson 99
Joint National Committee on Detection, Evaluation and Treatment of High Blood Pressure 154
Joint Reviewing Committee on Prescription Pricing 21
journals
  reviewing process 94
  pharmaceuticals financed 250-2

Lipid Research Clinics Coronary Primary Prevention Trial 153, 154
lisinopril-MSD 170

mass treatment *see* hypertension, hypercholesterolaemia, hormone replacement therapy
May and Baker (NZ) Ltd 25
medical education
  pharmaceutical industry input into 208-15, 257-8
  and sponsorship guidelines 222-3
medical practitioners
  fees of 27
  and prescribing practices 27-8, 42, 47-8
  and ethical code 222, 224, 258-9
  and pharmaceutical industry 223-7, 245-6, 282-3
  knowledge of drugs 246, 258, 290-1
  influence of advertising on 254, 260
  gifts and sponsorship 258-60
  see also general practitioners
medicalization 7, 181-7, 293
Medical Research Council
  and fenoterol 90
  and Depo-Provera 134
  and hypertension 150-1, 167
Medical Services Committee 31
Medicines Act 261, 279, 281
Medicines Adverse Reactions Committee 60
  fenoterol 70-73, 90-91, 114
  Depo-Provera 123, 126, 130, 135
Medicines Assessment Advisory Committee 59
Medicines Regulations 61, 65, 253, 261, 279, 281
medroxyprogesterone acetate *see* Depo-Provera
menopause 179, 186-92
  see also hormone replacement therapy
Merck Sharp and Dohme 22, 259
methyldopa 173
Mianserin 126

Mini-Gravigard 107-12
Multiple Risk Factor Intervention Trial 147

National Kidney Foundation 69
Nelson General Practice Prescribing Project 56, 262-3
New Ethicals 250-1
New Ethicals Catalogue 251
New Zealand Asthma Foundation 91
New Zealand Contraception and Health Study 127, 132-5
  Depo-Provera 132
New Zealand Ethical Pharmaceuticals Association *see* Researched Medicines Industry
New Zealand Formulary 20
New Zealand Pharmaceutical Manufacturers Association *see* Researched Medicines Industry
niacin 154
Norplant 138-9
Notes on prescription costs 28

oestrogen
  adverse effects of 183, 187, 193, 200
  oestradiol 187
  diethylstilboestrol 187
  conjugated equine oestrogen 187
  oestrogen deficiency disease 187, 189-92
  see also hormone replacement therapy
Ogen 192
orciprenaline
  adverse effects of 77-8
Ortho Pharmaceuticals 99
Oslo Study Diet and Anti-smoking Trial 153
osteoporosis 127-30, 191, 192, 194, 196-7

Patents Act 1953 25
patents, drug 24-5, 229-31
Patient Management 250
pelvic inflammatory disease (PID) 102-4
Pfizer 239
Pharmaceutical Advisory Committee 27
pharmaceutical benefits 18, 30-3, 40-4, 48-9
pharmaceutical industry
  history of 1, 24-33
  marketing by 2, 10, 54, 57, 59, 64, 93, 209-27, 245-85
  corporate practices of 5
  foreign investment in New Zealand 22
  international 65
  defence of products 93-4
  commercial, legal, ethical responsibilities of 218

journals 225, 250-2
acceptance of gifts 259
marketing techniques of
  talk-back radio and media 222, 253
  direct mailing 252-4
  journals 225, 250-2
  direct-to-consumer 274, 281-2
  sponsorship and gifts 208-15, 221-3, 223-5, 257-8
  sales representatives 248, 254-6
medical education 208-15, 221, 257-8
patient education 222
political pressure 12, 65, 74
prescribing information 249-56
regulation 5, 53-74, 114, 261-4
  voluntary regulation 269-85
  Britain 269-72
  United States 272-5
  Canada 275-6
  Australia 276-8
  Sweden 278-9
  New Zealand 279-84
representatives 254-6
research funding 216-18
tax avoidance 67
see also names of pharmaceutical companies; Research Medicines Industry
pharmaceutical legislation *see*
Control of Prices Act
Drug Tariff
Medicines Act
Medicines Regulations
Patents Act
Social Security Act
Social Security Regulations
Pharmaceutical Manufacturers Association *see* Researched Medicines Industry
Pharmaceutical Manufacturers Association of Canada
Code of Marketing Practices of 275-6
pharmaceuticals
  international consumption 3
  critical evaluation 4, 114
  policy 8, 288-93
  expenditure 9, 18, 30-3, 44, 53-74, 214
  regulation 10, 122, 125
  imports 21, 26, 64-5
  New Zealand manufacturing 21-2
  pricing 23-4, 29, 44, 289
  distribution channels 41-4
  clinical trials 67-70
  high cost medicines 67-73
  safety 126
  pharmacists' role 290
  prescribing variation 290-1
  prescribing formulary 291-2

public interest 292
social and psychological aspects
see also drugs; prescriptions; brand name drugs; generic drugs; drugs by specific brand or generic name
Pharmaceuticals Review Strategy 55-74
pharmacists
  contracts 28-30, 42-51, 290
  changing role 36-51
see also Chemists' Service Guild
Pharmaco 257
Pharmacology and Therapeutics Advisory Committee 28, 59, 68, 69, 73, 107
Pharmacy Authority 40
Pharmacy Board of New Zealand 19, 23
Pharmacy Plan Industrial Committee 19, 22, 28
Population Council 99, 100
Preferred Medicines Coordinating Centre 56
Premarin 185, 187, 192, 193, 194, 199
prescribing
  process 245-6
  information, professional 246-9, 261-3
  information, commercial 249-61
prescriptions
  charges 19, 22, 23, 45-8, 208
  brand substitution 235-42
progesteron
  adverse effects of 199
propranolol 151, 173
Public Health Association 57, 72

Researched Medicines Industry 10, 12, 26, 57, 61-5, 70, 219
  drug patents 229-32
  generic prescribing campaign 232-42, 260-1
  Code of advertising practice 253-4, 260, 270, 280-2
Retin-A 182
risk factors *see* hypertension, hypercholesterolaemia
RMI *see* Researched Medicines Industry
Roche 233, 274

Schering 257
Sex Hygiene and Birth Regulation Society *see* Family Planning Association
Smith, Kline and French 233, 235, 278, 281-2
Social Security Act 18, 21, 27
Social Security Fund 19, 21
Social Security Pharmaceutical (Supplies) Benefits 19
Social Security (Medical Benefits) Regulations 1941 27
Squibb 233, 253
Sterling Pharmaceuticals (NZ) Ltd 22

Stockholm Ischaemic Heart Disease
    Trial 153
Streptokinase 264

Therapeutic Notes 28, 248–9
TPA (tissue plasminogen activator) 264

Upjohn Company 119–41, 188, 199, 233, 259
  Depo-Provera 119–41
  Halcion 136
US National Women's Health
    Network 136, 138

Ventolin 281–2

Wholesale Druggists Federation 22
Wilson Foundation 188

women's health 113–15
  medicalization 181–7
  ageing 183–6
  menopause 179, 186–92
  cultural attitudes 189
  osteoporosis 192, 194–7
  cardiovascular disease 192, 197–200
  see also contraceptives; breast cancer;
    cervical cancer
World Health Organisation
  Collaborative Study of Neoplasia and
    Steroid Contraceptives 132, 140–1
  Clofibrate Trial 144, 152
  *Ethical Criteria for Medicinal Drug
    Promotion* 270, 277, 283
Wyeth-Ayerst Laboratories 274

Zocor
  marketing campaign 68–9